Ethical Assemblages of Artificial Intelligence

Helena Machado · Susana Silva

Ethical Assemblages of Artificial Intelligence

Controversies, Uncertainties, and Networks

Helena Machado 🆔
Institute of Social Sciences
University of Minho
Braga, Portugal

Iscte—University Institute of Lisbon,
CIES-Iscte
Lisbon, Portugal

Susana Silva 🆔
Department of Sociology
Institute of Social Sciences
University of Minho
Braga, Portugal

ISBN 978-981-96-4157-4   ISBN 978-981-96-4158-1   (eBook)
https://doi.org/10.1007/978-981-96-4158-1

This Palgrave Macmillan imprint is published by the registered company Springer Nature Singapore
Pte Ltd.
The registered company address is: 152 Beach Road, #21-01/04 Gateway East, Singapore 189721,
Singapore

If disposing of this product, please recycle the paper.

To our families and in loving memory of those we have lost.

Preface

In this book, we delve into the concept of ethical assemblages within artificial intelligence (AI). This idea captures the intricate interplay of technologies, people, organisations, values, ideas, and practices that shape our socio-material and symbolic landscapes. Focusing on AI in facial recognition and assisted reproductive technologies (ART), we explore how these ethical assemblages emerge in media, academic literature, and online discourses. Our approach highlights the dynamic encounters of controversies, uncertainties, and networks during ethical negotiation and contestation. Whether the reader is a researcher examining the evolving role of AI in society, a practitioner working with these technologies, or a policymaker concerned with their regulation, this book offers a critical examination of the controversies, uncertainties, and debates that often remain hidden from public discourse.

Our work in this book represents both a continuation and a transformation of our research over the past two decades. We departed from examining governance in security practices and reproductive health through emerging genetic technologies to exploring the social and ethical challenges of AI in these fields. Our focus remains on how advanced technologies that collect data from human bodies navigate the intersection of surveillance and care. We ask how can advanced technologies that gather data from human bodies fulfill various roles intersecting surveillance and care? Who determines the framings of surveillance and care, and who stands to benefit from them?

In the early 2000s, while pursuing our PhDs and later as early career researchers, we became captivated by the transformative potential of genetics in tackling significant societal challenges within Western societies, particularly in the realms of criminality and infertility. This period was marked by widespread enthusiasm for genetics, with substantial public and private investments directed towards the sequencing of the human genome project. Within this context, the criminal justice system experienced a revolutionary shift with the introduction of genetic databases and advanced technologies for suspect identification, heralding what became known as the genetic era. Concurrently, advancements in infertility treatments spurred the development of genetic mapping technologies and methods for embryo selection. In this context, over the past two decades, our research has delved into the intricate and context-dependent social dynamics surrounding these technologies.

One facet of our inquiry was centered on genetic and genomic advancements in forensic science applied to criminal identification. At that time, we have extensively investigated how genetic surveillance by states intersected with political processes that reframed collective issues as matters of security. Parallel to this, our research had focused on assisted reproductive technologies, particularly exploring attitudes towards donating gametes and embryos for research and human reproduction purposes. This line of inquiry sought to inform people-centered health policies and integrated care within socio-technical environments, while grappling with the societal and ethical dilemmas posed by disruptive technologies.

When reflecting on the intersection of these two lines of research, one aspect consistently stood out over the years, and we return to this in this book: the core ethical and governance concerns that prevailed in medical genetics and genomics were conspicuously absent in the field of forensics. In medical contexts, the emphasis was on protecting the autonomy and interests of patients. Conversely, in forensic contexts, the focus was on protecting the interests of the society as a whole. This illustrates a significant difference in how the balance between individual and public interests is framed in these two domains. This book recovers the tensions and blurred boundaries between individual and public interests by interrogating them, focusing on the particular cases of AI technologies applied to public security (facial recognition) and medicine (assisted reproductive technologies).

Two other prominent themes emerged from our past research which we continue to address in this book. First, there is an urgent need to develop ethical frameworks that truly reflect the relational nature of people and data. Our current ethical approaches need to evolve to better address the complex interplay between individual privacy and autonomy and the changing dynamics of the collective. Second, the importance of public engagement in the governance of AI applications in society remains a topic deserving more critical inquiry.

Our extensive journey has brought us to the forefront of contemporary issues explored in this book, centered around a new epoch marked by AI. Within this landscape, we delve into two distinct realms of AI applications. Firstly, we scrutinize the societal implications of AI-driven facial analysis, probing the intricate dynamics of power and politics surrounding the concept of "faciality" in today's societies. Secondly, our focus extends to the transformative impacts of technological and bioscientific innovations facilitated by AI, particularly within the realm of human reproduction, which are already shaping the future narratives of healthcare. Of particular concern, central to our past and current research efforts, is a crucial question: what fundamentally distinguishes and parallels the eras of genetics and AI? This inquiry is too intricate to be resolved in a single book or research project. Consequently, we anticipate that this question will continue to guide us and our research teams in the years to come.

This interrogation has led us to explore not only the trajectories that define the scientific and technological revolutions in both genetics and AI, but also the profound ethical questions and societal concerns that accompany these advances. The rapid pace of innovation in both fields introduces uncertainties that underscore the need for continued research, ethical deliberation and adaptive policy-making. Furthermore, both genetics and AI have been extensively explored in speculative fiction, which has significantly influenced public perceptions and sparked ethical debates long before many of these technologies became feasible. This intersection between speculative fiction and the realisation of acceptable and viable technologies has prompted us to delve deeper into the crafting of media, scientific and online discourses that bridge the imagined and the actual, shaping our understanding and governance of these revolutionary fields.

The rapid pace and uncertainty, and the role of speculation and futurism, lead us to ask critical questions in this book about the implications of the AI advances: how do new concepts of public security and health care take shape and spread through AI innovations? What role do social values and ethical considerations play in shaping current and future preferences for AI-mediated security and healthcare solutions? Importantly, we examine the extent to which these visions of the present and future are subject to debate and contestation, configuring multiple versions of AI technologies.

Braga and Lisbon, Portugal   Helena Machado
                                                        Susana Silva

Acknowledgments

In writing this book, we have drawn upon the invaluable support and inspiration extended by family, colleagues, students, and institutions, whose contributions we gratefully acknowledge. In June 2023, we established the AIDA (Artifical Intelligence, Data & Algorithms) Social Sciences Research Network at the Institute of Social Sciences, University of Minho, Portugal. This dynamic community has fostered an enriching and stimulating environment, uniting researchers across different generations, and has been instrumental in nurturing the intellectual, personal, and emotional dimensions underpinning the ideas explored in this book. We offer special appreciation to Rafaela Granja, Susana de Noronha, and João Sarmento for their insightful critique of early versions of this manuscript. Furthermore, we are deeply grateful for the vibrant seminars organized within the framework of AIDA, which have brought together colleagues from diverse disciplines and countries. We acknowledge with gratitude the teachings and suggestions imparted by fellow AIDA members: Emília Araújo, Sheila Khan, Laura Neiva, Filipa Queirós, and Maria João Vaz. A special thank you goes to Marion Duval, whose insightful feedback have greatly enhanced the clarity and quality of our work. We are also deeply grateful to the anonymous reviewers, whose thoughtful critiques and constructive suggestions have significantly shaped the development of this book.

We deeply appreciate the support and inspiration of our colleagues, whose contributions, often unheralded, have been immeasurable. The multinational and interdisciplinary consortium "SolPan—Solidarity in Times of a Pandemic," under the leadership of Barbara Prainsack at the University of Vienna, brought together researchers from ten European countries. Similarly, SolPan+, involving seven Latin American countries, enriched our intellectual landscape with vibrant debates and exchanges of ideas. Interacting with researchers from diverse social, cultural, and political backgrounds provided invaluable insights into the complexities of our world—a perspective that was integral to shaping the themes explored in this book, emphasizing diversity, heterogeneity, and the nuanced interplay of ideas.

While our individual journeys often led us down separate paths, we have been fortunate to embrace the opportunities that academic life has provided for ongoing collaboration. Moreover, we have been deeply enriched by the support of many colleagues and research grants, each deserving of recognition. Regrettably, it is not feasible to acknowledge everyone individually.

As a founding member of the Forensic Databases Advisory Board appointed by the International Society of Forensic Genetics (2021–2023), Helena has benefited greatly from the lively debates about the risks, harms and benefits of collecting genetic samples for forensic purposes. The intense and thought-provoking conversations with colleagues from both forensic genetics and law (Maria Eugenia D'Amato, Nathan Scudder, Martin Zieger, and Yann Joly) have profoundly challenged prevailing notions of privacy, liberties and human rights, as well as the delicate balance between surveillance and care. Helena has also received significant support through several research grants, notably from the European Research Council (Consolidator Grant No. 648608, 2015–2021) and the Portuguese Foundation for Science and Technology (Grant No. 01-0124-FEDER-009231; IF/147549; 2419-10364/13-7, awarded between 2010–2017). The research undertaken in preparation for the European Research Council Advanced Grant entitled "Facial

Recognition Technologies. Etho-Assemblages and Alternative Futures (fAIces)" (Grant No. 101140664), along with the project's initial stages, played a crucial role in shaping this book, making it an early outcome of fAIces and influencing its future development.

As a member of the "Biosocial Birth Cohort Research Network" (2019–2024), coordinated by Sahra Gibbon at University College London (Wellcome Trust Small Grant reference 218187/Z/19/Z), Susana was inspired by the unique opportunity for interdisciplinary exchange, reflection and critique among a community of international researchers from a wide range of life and social science disciplines to explore the social, ethical and methodological challenges and opportunities of developing biosocial science. Susana has also received significant support through several research grants, notably from the Portuguese Foundation for Science and Technology (Grant Nos. IF/01674/2015; POCI-01-0145-FEDER-016762; IF/00956/2013; FCOMP-01-0124-FEDER-019902; FCOMP-01-0124-FEDER-014453, awarded between 2011–2021).

Other colleagues and friends have continually provoked us with critical reflections on our research. We would like to thank in particular António Amorim (in memoriam), Mariana Amorim, Sílvia Fraga, Cláudia de Freitas, Catarina Samorinha, Joëlle Vailly, Nina Amelung, and Tiago Santos Pereira.

Finally, we would like to express our heartfelt gratitude to our students, friends, family and all those who have contributed to our research endeavours. Their unwavering support, creative sparks and insightful critiques have been woven into the fabric of this book. It stands not only as a testament to our collaborative efforts, but also as a celebration of the curiosity and ingenuity that drive our shared quest to understand social complexity.

About This Book

This book embarks on a journey into the ethical assemblages of artificial intelligence (AI). Focusing on facial recognition technologies and assisted reproductive technologies (ART), our research uncovers the intricate and interconnected socio-material and symbolic processes where diverse elements—technologies, people, organisations, values, ideas, and practices—interact and engage in negotiations that shape evolving ethical assemblages.

From the widespread integration of facial recognition in daily life to the burgeoning potential of AI in enhancing healthcare in the field of ART, each chapter illuminates the controversies and uncertainties surrounding these innovations. We pose critical questions about visibility and public discourse: why do some ethical debates around AI gain widespread attention, while others remain obscured? How do different social actors involved in dynamic and intricate networks of social relations and practices align or diverge in their ethical stances?

Through the lens of assemblage theory, we navigate the heterogeneous and dynamic nature of AI's ethical challenges. We uncover the controversies, uncertainties, and networks that underpin AI's ethical landscapes, exploring resistance, refusal, and dissensus, along with resignation and acceptance, in shaping ethical governance and societal norms. This book is not just a scholarly endeavor but a call to action for inclusive and informed dialogues on the ethical dimensions of AI in shaping our collective future.

Contents

About the Authors

Helena Machado Full Professor of Sociology at University of Minho (Braga, Portugal) and Coordinator Researcher at Iscte-University Institute Lisbon, CIES-Iscte (Lisbon, Portugal). Her research lies at the intersection of sociology, science and technology studies, critical surveillance studies, and criminology, focusing on emerging technologies for human identification and behaviour prediction. She specialises in facial recognition and forensic genetics and genomics, exploring their implications for governmentality and subjectivities. Her work also examines the societal and political impacts of artificial intelligence, particularly narratives around surveillance and care, and public involvement. She has been awarded prestigious grants, including from the European Research Council and the Portuguese Foundation for Science and Technology. She is the founding member and coordinator of the AIDA (Artificial Intelligence, Data & Algorithms) Social Sciences Network.

Susana Silva Associate Professor of Sociology at University of Minho (Braga, Portugal) and Researcher at Centre for Research in Anthropology

(CRIA-UMinho). She has written about people-centered health policies and integrated care in socio-technical environments, with a focus on the societal and ethical implications of reproductive and genetic technologies, critiquing their impact on research, health regulation, and clinical practices. Currently, she investigates pragmatic ethics, ethics of care, and public engagement in AI regulation, especially in medicine and forensics. Her work addresses emerging ethical challenges, particularly in AI applications for human reproduction, aiming to promote informed and responsible practices in these evolving fields. She is member of the AIDA (Artifical Intelligence, Data & Algorithms) Social Sciences Network.

List of Figures

List of Tables

1

Ethical Assemblages of Artificial Intelligence

Abstract This introductory chapter explores the concept of ethical assemblages within artificial intelligence. The term ethical assemblages refers to the complex interplay of technologies, people, organisations, values, ideas, and practices that shape socio-material and symbolic landscapes through ethical negotiations. We propose a conceptual framework that sustains the examination of the real-world cases of facial recognition technologies and assisted reproductive technologies in the next chapters, by exploring the dynamic processes of ethical contestation, uncertainty, and controversy that emerge in the media, academic literature, and online discourse. Through the lens of assemblage theory, we emphasise the fluidity and interconnectedness of ethical dilemmas, highlighting how different social actors—including technologists, medical professionals, policy makers, academics, and the public—navigate and influence ethical frameworks. We also incorporate intersectional theory to explore how multiple axes of difference, such as gender, race, and class, intersect within these assemblages. By analysing the complex networks and power dynamics at play, ethical assemblages provide a critical understanding of how ethical norms and practices are evolving in response to AI innovations, and offer insights into the broader implications for society and governance.

© The Author(s), under exclusive license to Springer Nature Singapore Pte Ltd. 2025
H. Machado and S. Silva, *Ethical Assemblages of Artificial Intelligence*,
https://doi.org/10.1007/978-981-96-4158-1_1

1

Keywords Ethical assemblages · Artificial intelligence · AI ethics ·
Facial recognition technologies · Assisted reproductive technologies

Introduction

In this book, we delve into the concept of ethical assemblages within
the realm of artificial intelligence (AI). This term captures the intricate
and interconnected social processes where diverse elements—technologies, people, organisations, values, ideas, and practices—interact and
engage in ethical negotiations and discussions, thereby (re)shaping socio-
material and symbolic landscapes. Our focus is on ethical assemblages
related to AI in facial recognition technologies and assisted reproductive
technologies (ART), as they emerge in media sources, expert academic
literature, and online sites promoting or opposing these technologies.
Ethical assemblages bring together people, technology, perceptions of the
social world, and meanings. Our approach in this book is to understand these assemblages as encounters of controversies, uncertainties,
and networks, combining durability and contingency, homogeneity and
heterogeneity, contestation and connection (Aradau & Blanke, 2022,
p. 12).

Among the pivotal questions explored within this book are: which
ethical debates in the AI domain gain prominence, and which ones
remain overlooked or hidden? Which social actors, institutions, and
organisations align with each other, and which ones diverge when
engaging with ethical controversies and uncertainties? What complex
networks shape these processes, and what narratives and practices are
positioned within, outside, or on the fringes?

A provocative examination of the particular real-world cases of facial
recognition technologies and ART will ensure a tangible understanding
of ethical assemblages. The rationale for selecting these cases lies in three
main reasons. First, a temporal contrast. While facial recognition technologies have been in expansive use for the past two decades, AI in ART
represents a more recent entrant. This coincidence makes it possible to
dissect how ethical controversies and uncertainties interact with different
stages of an innovation cycle—from conception and development to
adaptation and redirection. Second, a contrasting public visibility. While

facial recognition technologies have a widely widespread usage in society, the use of AI in ART tends to circulate in less visible spaces such as professional societies in the field of reproductive medicine and fertility, scientific journals, and clinic websites. These circuits of variable and differentiated visibility allow us to explore how ethical controversies and uncertainties are intertwined with a variety of social actors and domains that form distinct yet interwoven networks. Finally, contrasting sites and expressions of contestation. Facial recognition technologies will be explored in this book as an exemplary case of convergence between ethical concerns, public controversies and uncertainties, and social movements of refusal, resistance, and dissensus, while AI in ART will illustrate a case dominated by expert and technocratic approaches to controversies and uncertainties, professional dissensus and refusal, and shared engagement with present and future ethical concerns.

Before presenting the conceptual framework for ethical assemblages in AI, we invite the reader to join a woman—let's call her Maya— on a multifaceted journey into AI. Imagine a scenario where she visits a fertility clinic renowned for its pioneering use of advanced AI technologies, promising the most personalized and successful treatment possible.

Upon arrival, Maya is admitted by having her face scanned and her authorisation confirmed by the security system. This process evokes memories of the initial challenges she faced when registering as a client at the clinic. The first step was to upload a photo, but technical difficulties arose. "Our apologies", the Integrated Clinic System (ICS) staff had said. "For some reason, the system incorrectly flags photos of black or brown people as errors".

Maya is an activist advocating for the strict regulation of facial recognition technology. She experiences mixed feelings, oscillating between protesting against the technology that scans her face and seeking assistance from AI in her infertility treatment, specifically through the matching of an egg donor to herself via facial recognition software.

As Maya sat in the clinic's waiting room, her hands clasped tightly around a printed pamphlet entitled "Artificial Intelligence will help you fulfil your dreams of parenthood". Despite having read it countless times, the words still held a complex mixture of hope and trepidation. Today marked a pivotal moment in her journey through ART.

Her heartbeat accelerates as she noticed the discreet cameras positioned around the room. These were part of the clinic's advanced facial recognition system, designed to monitor the emotional state of patients. The AI analysed micro-expressions, measured stress and anxiety levels, and sent the data directly to the attending doctors. This system promised to facilitate a more empathetic approach to treatment, but the intrusion into her private emotions left Maya feeling exposed.

When she was called into the fertility specialist's office, Maya took a deep breath. The doctor greeted her warmly, but his eyes flickered to the screen displaying Maya's emotional state. "I can see you're feeling a little anxious today", he said gently. "That's perfectly normal". Maya nodded, her thoughts racing back to the second, more personal aspect of the AI in her treatment. The same technology that assessed her stress was also used to match her phenotype with an egg donor and to assess and select the most competent eggs and sperm with speed and precision and the best embryos for transfer. The AI system was able to analyse her genetic make-up, physical characteristics, and even ancestral background to predict the best match for successful implantation and the future birth of a healthy child.

"There are many uncertainties in this process", the doctor explained, sensing Maya's unease. "We are utilising AI to help us in selecting the optimal strategy for embryo transfer, but it is not without risks". Maya was aware of the ethical debates surrounding the use of AI technologies to select embryos. The idea of an AI making decisions about potential life was both wondrous and unsettling. She had come across stories of AI making mistakes, and the possibility that these errors might impact her future child was a constant source of concern. As the consultation proceeded, the doctor provided reassurance to Maya regarding the rigorous ethical standards and oversight in place. "Our AI system is designed to work under the careful supervision of the best experts to ensure that every decision is carefully considered by our team and tailored to your personal circumstances".

As she left the clinic, Maya felt a mixture of relief and apprehension. The uncertainties surrounding AI were complex, yet they also held the promise of a future she desperately wanted. As she walked down the busy street, the city's facial recognition cameras silently tracked her movements, a tangible reminder of how intertwined her life had become with these invisible technologies.

Despite her mixed feelings, Maya felt a renewed sense of determination. As an activist protesting against the widespread use of facial recognition technology in public spaces, she had a deep understanding of the power and dangers of AI. Her personal journey through ART had shown her the double-edged nature of these technologies. She knew that her struggle was not merely about the strict regulation of a technology that could potentially threaten civil liberties; it was about ensuring that AI served the best interests of humanity while protecting individual rights and dignity. Maya felt she was part of a larger story, one that combined human aspirations with the cutting edge of AI, navigating through controversy, uncertainty, and hope. She decided to persist in her advocacy, leveraging her experience to advocate for a future where technology coexists with humanity, justice, and equality.

Ubiquity, Speculation, (In)Visibility, and Contestation

The introductory fictional story features two very different applications of AI—facial recognition technologies and ART—and it is difficult to distinguish between technologies that are currently available in the real world and those that remain speculative. Facial recognition technology is now widely used, even described as ubiquitous, in security systems, law enforcement, smartphone unlocking, and social media platforms, while AI in ART is still in the early stages of clinical applications to improve well-known techniques such as in vitro fertilisation (IVF) and egg/sperm donation or embryo transfer. Speculative advances in both areas could include AI predicting behaviour and emotions through facial recognition, or selecting the best embryos and gametes developed in vitro, potentially allowing future children to be conceived through an AI selection process.

Proponents argue that AI applications for facial recognition technologies and ART can significantly improve safety and health outcomes. They highlight the potential benefits of these innovations, which include enhanced public security, improved ART success rates and personalised healthcare. These technologies serve as an interface that enables translation between current societal problems, such as insecurity and delayed childbearing, and visions of the future.

However, the social and ethical implications of AI technologies are both substantial and multifaceted. While some implications are highly visible and widely discussed, others are less noticeable or hidden, drawing varying levels of media and public attention. The element of visibility is crucial for examining how controversies and uncertainties surrounding facial recognition technologies and ART circulate in society and are influenced by different social actors and their interactions. For example, facial recognition technologies may operate covertly in public spaces, prompting widespread media coverage, public attention and opposition from activist movements, and raising concerns about privacy and surveillance that involve not only technology companies and privacy advocates, but also the general public and legal authorities. The uses of AI in ART tend to occur in private spaces, triggering debates in bioethics and medical forums, although they can also be hidden within private clinical practices, and raising ethical dilemmas about the limits of intervention in human reproduction, access and inequality, informed decision-making, effectiveness, and clinical validation, inviting opinions and judgements from medical professionals, bioethicists, and society at large. AI technologies frequently give rise to debates regarding controversies and uncertainties that encompass not only developers, users, and regulators but also a broader array of social actors involved in dynamic and intricate networks of social relations and practices. These complex networks illustrate how debates about AI technologies are not confined to technical or ethical domains, but are deeply embedded in broader social contexts where different forms of power, knowledge and contestation encounter.

In order to illuminate the multiple ways in which different social actors navigate, intersect and contest the evolving landscape of emerging

AI technologies, we will follow the methodological framework articulated by Aradau and Blanke (2022, p. 12), known as the "scene". This approach involves the analysis of contentious and uncertain scenarios, including scenes of controversies, debates, disputes, and even scandals surrounding AI technologies (Aradau & Blanke, 2022, p. 7).

Heterogeneity, Processuality, and Intersectionality

Assemblage theory is used in this book to highlight the fluidity and interconnectedness of social complexity. By adopting the flexible conceptual tool of assemblage theory, we aim to foster inductive theory building and epistemological openness, thereby enabling us to explore the ontological diversity of agency within complex networks of technologies, people, organisations, values, ideas, and practices that co-constitute controversies and uncertainties around AI technologies. The assemblage framework thus advocates for "the rejection of unity in favour of multiplicity" (Nail, 2017, p. 22), highlighting the intricate interplay of diverse components.

While the notion of assemblage has increasingly emerged in various fields, including philosophy (DeLanda, 2002, 2016; Deleuze & Guattari, 1987), social anthropological studies (Rabinow, 2003), and gender studies (Puar, 2007), it has become particularly evident in the study of science and technology in the context of Bruno Latour and Michel Callon's contributions to actor-network theory (Latour, 2005; see also Müller, 2015). Deleuze and Guattari's emphasis on heterogeneity underscores the need to consider AI technologies not in isolation, but as part of broader, continuously evolving networks of relations. This perspective aligns with Bruno Latour's actor-network theory, which sees the social as "a process of associations between heterogeneous elements" (Latour, 2005, p. 1), including human and non-human actors. Latour's processual terms such as "the assembled", "assembling", and "reassembling" (Latour, 2005, p. 1) capture the fluid and transient nature of these associations, reflecting the temporary stabilisation of ethical and social norms.

In our perspective, heterogeneity and processuality are two core elements of assemblage theory that provide a compelling framework for exploring the ethical dilemmas surrounding facial recognition and ART. In the context of these technologies, heterogeneity refers to the diverse elements that constitute the assemblage, including technological, biological, ethical, and social elements. Processuality implies the evolving nature of ethical norms and social practices, which are continuously shaped by ongoing interactions and technological developments, generating new societal impacts that require a flexible and responsive approach assessing them.

The study of ethical assemblages also benefits from context-sensitive intersectional theory. The term "intersectionality" is usually attributed to Kimberlé Crenshaw, a legal scholar, who coined the term in 1989. In this context, intersectionality emphasizes the interconnected nature of social identities, such as race, gender, and class, and the ways in which these identities intersect to shape individuals' experiences of oppression and privilege. While Crenshaw's work is foundational, the concept of intersectionality has deep roots in grassroots women's movements in the Global South, where radical critiques challenged the dominance of white, middle-class women's perspectives on global issues. These movements, alongside the Black women's movement in the United States, laid the groundwork for the development of intersectionality as an analytical tool (Bastia et al., 2022). Intersectionality helps us to understand how overlapping identities—such as race, gender, and class—shape the experiences and ethical concerns surrounding facial recognition and ART, highlighting the complexity of addressing issues from a singular, universal standpoint.

This diversity, highlighted through both intersectionality and assemblage theory, challenges the assumption of a singular, unified approach to ethical issues. Instead, it advocates for context-sensitive ethical frameworks that consider the potentially conflicting interests at play and the diverse and evolving nature of the elements involved, from technological capabilities and medical practices to legal regulations and societal values. By integrating intersectionality, we are reminded of the importance of accounting for the intersecting identities and power dynamics that influence how these technologies impact different groups.

The intersectional approach calls for the consideration of multiple axes of difference and hierarchies (gender, race, class, economic and social capital, etc.) within assemblages. This approach necessitates the exploration of how different social interpretations and practices emerge from the same set of events, shaped by different social actors (Amelina, 2017, 2021). By examining these "axes of difference" through the lens of assemblage theory, we can analyse facial recognition and ART with a more focused attention on the nexus of domain-specific hierarchies. These include scientific authority in science, capital accumulation in technology corporations, collective decision-making in politics, contestation within civil society organisations or professional societies, and individual preferences of ordinary people, among others. This is crucial for tracing the interplay of diverse logics of actors from each domain in the processes of ethical assemblages embedded in networks and arranged in particular manners.

The assemblage theoretical perspective does not assume that the structures of different domains are homologous or of a similar nature. Therefore, when attempting to reconstruct complex and potentially contradictory hierarchies within domains, it is of the utmost importance to use a very careful analytical conceptualisation in order to fully understand their malleable and hybrid compositions (Fig. 1.1).

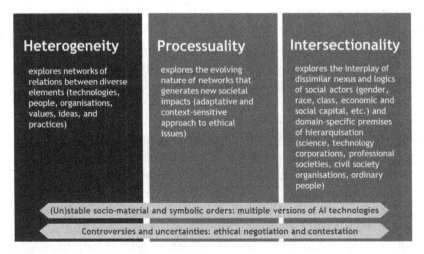

Fig. 1.1 Conceptual framework for ethical assemblages in artificial intelligence

Controversies, Uncertainties, and Networks

This book delves into how controversies and uncertainties create a complex and dynamic landscape in which contestation and public debates about the meanings and ethical implications of these technologies emerge, evolve, and are attempted to be resolved through social and technological rearrangements. It argues that controversies and uncertainties are integral to the fabric of technological innovation and its governance. Ethical assemblages are collectively constituted through evolving relationships within specific networks that bring together diverse elements, including multiple social actors and domain-specific premises, each with its own logics and nexus of hierarchisation, social values, political interests, and power relations (Bigo et al., 2019). These networks provide a temporary, fluid, and shifting collective sense and sense-making of what might be considered "ethical" and "visible", (re)shaping socio-material and symbolic landscapes. Constant reconfigurations of the relationships between academic scientists, technology companies, states, regulators, professional societies, civil society organisations, and people open space to multiple versions of AI technologies. Agency is understood as an effect of the relationalities that constitute the network's own thread and path, rather than the result of individual efforts.

Science and Technology Studies (STS) insights into controversies allow us to analyse the logics, social values, interests, and power relations that drive the opening and closing of controversies and the (in)visibility of uncertainties around facial recognition technologies and ART within specific networks. Beyond the initial focus on factual disputes primarily involving competing expert communities (Jasanoff et al., 1995; Pinch, 2015), historical and field studies of controversies over scientific issues and technological innovations had established the distinctive STS claim that the formulation of knowledge and the organisation of political interests tend to go hand in hand (Collins & Pinch, 1998; Hagendijk & Meeus, 1993; Venturini, 2010). For our case studies, this is crucial for understanding academia's uneasy relationships with private firms and industry, and the socio-material and symbolic orders that underlie them, among other interactions that are currently under critical scrutiny. It is our contention that STS controversy studies should occupy a pivotal and enduring inspiring role in the analysis of AI, ethics, and society. This is not only to address persistent concerns about bias, discrimination, and black-box systems, but also to facilitate a deeper understanding of the politics of knowledge. Furthermore, it offers a distinctive approach to studying the partiality of knowledge and the corresponding multiple versions of AI technologies (Marres, 2015; Venturini, 2012).

We will also explore how scenes of controversy and uncertainty link different forms of contestation, from dissensus to refusal and resistance (Aradau & Blanke, 2022, p. 209), along with resignation and acceptance as responses that differ from but complement resistance, refusal, and dissensus in terms of how individuals and groups engage with AI. Resistance describes opposition to direct domination, while refusal encompasses disavowals, rejections, and manoeuvres with and away from diffuse and mediated forms of power (Prasse-Freeman, 2022). Resistance and refusal thus operate in a quasi-dialectical relationship (Prasse-Freeman, 2022), meaning that through a play of recursivity between seemingly opposite strategies (direct confrontation versus evasion), social actors come to fortify stronger positions from which to persevere. Dissensus, defined as the active expression of disagreement and dissent against established norms and practices, further enriches the dynamic of contestation by highlighting the disruption of consensual politics and hegemonic frameworks, and claiming for the reconfiguration of power relations.

Together, resistance, refusal, and dissensus serve to both constitute the conditions for one another's existence, thereby enhancing the particular interventions and claims each makes; additionally, they reveal multiple and fluid ways in which social actors and domain-specific premises oscillate between direct confrontation, governmental navigation, and active dissent in the context of negotiating AI ethics. These dynamics coexist within networks and continually redefine and challenge the contours of power and practices in the field of facial recognition and ART, illustrating how ethical debates about AI technologies are deeply embedded in broader social contexts where different forms of power, knowledge, and contestation interact.

References

Amelina, A. (2017). *Transnationalizing inequalities in Europe: Sociocultural boundaries, assemblages, regimes of intersection*. Routledge.

Amelina, A. (2021). Theorizing large-scale societal relations through the conceptual lens of cross-border assemblages. *Current Sociology, 69*(3), 352–371. https://doi.org/10.1177/0011392120931145

Aradau, C., & Blanke, T. (2022). Algorithmic reason: The new government of self and others. *Oxford University Press*. https://doi.org/10.1093/oso/978019 2859624.001.0001

Bastia, T., Datta, K., Hujo, K., Piper, N., & Walsham, M. (2022). Reflections on intersectionality: A journey through the worlds of migration research, policy and advocacy. *Gender, Place & Culture, 30*(3), 460–483. https://doi.org/10.1080/0966369X.2022.2126826

Bigo, D., Isin, E., & Ruppert, E. (2019). *Data politics: Worlds, subjects, rights*. Routledge

Collins, H., & Pinch, T. (1998). *The Golem: What you should know about science*. Cambridge University Press.

DeLanda, M. (2002). *Intensive science and virtual philosophy*. Continuum.

DeLanda, M. (2016). *Assemblage theory*. Edinburgh University Press

Deleuze, G., & Guattari, F. (1987). *A thousand plateaus: Capitalism and schizophrenia*. University of Minnesota Press.

Hagendijk, R., & Meeus, J. (1993). Blind faith: Fact, fiction and fraud in public controversy over science. *Public Understanding of Science, 2*(4), 391–415. https://doi.org/10.1088/0963-6625/2/4/008

Jasanoff, S., Markle, G., Petersen, J., & Pinch, T. (eds.) (1995). *Handbook of science and technology studies*. Sage Publications

Latour, B. (2005). *Reassembling the social: An introduction to actor-network theory*. Oxford University Press.

Marres, N. (2015). Why map issues? On controversy analysis as a digital method. *Science, Technology & Human Values, 40*(5), 655–686. https://doi.org/10.1177/0162243915574602

Müller, M. (2015). Assemblages and actor-networks: Rethinking socio-material power, politics and space. *Geography Compass, 9*(1), 27–41. https://doi.org/10.1111/gec3.12192

Nail, T. (2017). What is an assemblage? *SubStance, 46*(1), 21–37. https://doi.org/10.1353/sub.2017.0001

Pinch, T. (2015). Scientific controversies. In J. D. Wright (Ed.), *International encyclopedia of the social & behavioral sciences* (2nd ed., pp. 281–286). Elsevier.

Puar, J. (2007). *Terrorist assemblages: Homonationalism in queer times.* Duke University Press.

Prasse-Freeman, E. (2022). Resistance/refusal: Politics of manoeuvre under diffuse regimes of governmentality. *Anthropological Theory, 22*(1), 102–127. https://doi.org/10.1177/1463499620940218

Rabinow, P. (2003). *Anthropos today: Reflections on modern equipment.* Princeton University Press.

Venturini, T. (2012). Building on faults: How to represent controversies with digital methods. *Public Understanding of Science, 21*(7), 796–812. https://doi.org/10.1177/0963662510387558

Venturini, T. (2010). Diving in magma: How to explore controversies with actor-network theory. *Public Understanding of Science, 19*(3), 258–273. https://doi.org/10.1177/0963662509102694

2

Navigating the Complex Landscape of Facial Recognition Technologies

Abstract This chapter explores the development, applications, and ethical dilemmas of facial recognition technologies. It begins by describing how these technologies work, from traditional methods to advanced deep learning techniques, and traces the history of the technologies from their military origins to their widespread use in security, consumer devices, and beyond. We critically examine the ethical concerns surrounding facial recognition technologies, including issues of privacy, bias, discrimination, and the normalisation of surveillance. By doing so, we challenge simplistic views of these technologies as inherently good or bad, arguing instead for a nuanced understanding of ethical dilemmas as relational and contextual. The chapter also discusses the global implications of facial recognition technologies, particularly in contexts of power and social control. Through the lens of assemblage theory, it explores how ethical understandings of facial recognition technologies emerge, evolve, and influence societal values, highlighting the complex interplay between technology, ethics, and power.

Keywords Facial recognition technologies · Ethical assemblages · Artificial intelligence · Ethical dilemmas · Surveillance · Social control

Introduction

Facial recognition technologies, one of the most widely used artificial intelligence (AI) technologies, typically work by analysing facial features extracted from images or videos, including the distance between the eyes, the shape of the nose and the contours of the face.[1] Primarily, these technologies serve two main purposes: authentication (is this person who they claim to be?) and identification (who is this person?). However, these technologies are also being developed to support the assessment of emotional states, the inference of behavioural tendencies and personality traits, and the estimation of demographic characteristics such as age, gender, and ethnicity (Hupont et al., 2022; Kaur et al., 2020). Kelly Gates (2011, p. 152) suggests that facial recognition technologies can be divided into two main types: facial identification (associating the image of a face with a specific person) and facial analysis (extracting information from a facial image such as race, gender, age, emotion, sexual orientation, or even propensity to commit a crime).

From seemingly innocuous uses like tagging photos on social media to questionable applications such as racial or sexual profiling, facial recognition technologies are becoming increasingly ubiquitous and are much part of everyday life. However, the historical evolution of facial recognition technologies unveils a multifaceted journey spanning from their inception in national security contexts to their widespread integration into everyday life, accompanied by complex ethical considerations and societal impacts.

The objective of this chapter is to challenge the oversimplified perceptions of "good versus evil" in the realm of facial recognition technologies. Instead, we examine the intricate facets surrounding these technologies, including the diverse ethical dilemmas they present. To do this, we follow the many disparate links that socially construct such ethical dilemmas articulated about and within facial recognition technologies. Our aim

---

[1] Facial recognition technology is a subset of computer vision, itself a form of artificial intelligence. It is not a single technology, but rather an umbrella term for a number of technologies that provide facial identification and/or analysis capabilities. We prefer to refer to technologies (plural) rather than the singular because there is no single, unified, or universal system of facial recognition.

is to present an epistemological position that explicitly rejects simple binaries, such as beliefs that the use of facial recognition technologies is inherently negative and that opposition to them is inherently positive. We contend that ethical dilemmas should be addressed as social practices and as modes of creating meanings, which are always relational. The question of whether facial recognition technologies are framed as positive or negative is contingent upon one's position within the larger sociotechnical systems. These visions give rise to power relationships that are in a state of constant flux, and which manifest in larger assemblages of material, social, and symbolic relationships that extend beyond the framings of ethical challenges themselves.

To follow ethical dilemmas of facial recognition technologies we employ the methodological framework articulated by Aradau and Blanke (2022, p. 12), known as the "scene". This approach entails dissecting contentious scenarios, encompassing controversies, debates, disputes, and even scandals surrounding facial recognition technologies. In doing so, rather than simply following connections of disparate elements, we explore assemblages that include the social, political, and economic glue that holds those connections in place long enough to yield social truths or scientific facts.

The first section of this chapter defines facial recognition technologies and explains how they work. The second section outlines a brief history of facial recognition, tracing its origins from military-supported laboratory research to its widespread use in national security and consumer devices. The third section addresses ethical challenges posed by these technologies, contrasting the perspectives of various social actors (developers, activists, and critical scholars) on their implications. The fourth section delves into significant controversies and uncertainties surrounding facial recognition, such as mass surveillance, bias and errors, oppression and discrimination of minorities, and public and private partnerships. In the fifth section of this chapter, we highlight instances of resistance towards facial recognition technologies that arise in specific social and political contexts within various countries. The sixth section examines ethical assemblages, understood as unstable and evolving understandings of what is ethical and unethical, which emerge and take shape through networks involving a variety of social actors, including academic scientists, technology companies, governments, regulators, non-governmental organisations, and the general public.

The concluding section presents a summary of our approach to ethical assemblages, networks, uncertainties and controversies, with a particular focus on the emerging effects of diverse and complex societal and technological rearrangements collectively constituted rather than resulting from the actions of single actors (Konrad, 2006). Assemblage theory provides a valuable framework for understanding the complex and dynamic processes that shape social and symbolic configurations. These configurations, which are both processual and structural, merge diverse processes and elements into heterogeneous and relational networks of social actors, institutions, and social values. These networks collectively contribute to the creation, sharing, and legitimisation of knowledge and meaningmaking. Simultaneously, we recognize ethical components as having diverse roles, producing varied effects, and evolving over time, acknowledging the different and sometimes conflicting interests involved.

How Facial Recognition Technologies Work

Facial recognition technologies can be broadly categorised into two main types: traditional facial recognition and deep learning-based facial recognition. Traditional facial recognition, which is widely used in security systems and law enforcement, compares facial features in an image or video with a known database in order to identify individuals. In contrast, deep learning-based facial recognition employs AI techniques typically to learn facial features directly from extensive datasets, which may enable better performance under various conditions such as different lighting or angles. As Greenfield (2017) and Kaur et al. (2020) have observed, the development of deep neural networks and machine-learning technologies has significantly enhanced the capabilities of facial recognition algorithms. Previously, the complexities of the human face posed significant challenges to computer vision. However, recent advances in AI have enabled algorithms to recognise visual patterns and adapt to change. The advent of machine-learning algorithms has been a pivotal factor in the advancement of facial recognition technologies.

In contrast to rule-based algorithms, which rely on pre-defined instructions, machine-learning algorithms are capable of generating their own rules through a training process (Fry, 2018). Initially, the algorithm is presented with a vast collection of facial images. This dataset must be "labelled", which means that it must already include the pertinent information that the algorithm aims to learn, such as the gender of the individuals depicted. With each image it encounters, the algorithm attempts to ascertain the gender of the person depicted and then cross-references it with the provided information. Through a process of trial and error, coupled with an optimisation process grounded in statistical analysis, the algorithm is expected to incrementally refine its parameters to enhance its accuracy.

Bakir and McStay (2022, p. 8) argue that while the term "optimization" initially implies efficiency and the maximisation of effectiveness, its implications extend into political and contentious realms. This raises questions regarding the determination of what is optimal, the beneficiaries and victims of optimisation decisions, and the sacrifices made in the process. As the algorithm processes thousands, if not millions, of images, it develops its own guidelines for facial recognition and analysis (Fry, 2018, p. 11).

It is of paramount importance to consider the nature of the reality and truth that these algorithms produce. Louise Amoore examines the type of truth asserted by algorithms when they engage in facial identification or analysis. For example, she demonstrates how a facial recognition algorithm employed in urban policing can identify faces due to its exposure to a "labelled or unlabeled" set of training data from which the algorithm generates its model of the world (Amoore, 2020, p. 136). Amoore elucidates that as deep learning algorithms gradually construct their own interpretations from unprocessed, unlabelled data, they establish patterns of normalcy or deviation within the data. Consequently, the veracity of machine-learning algorithms does not inherently negate the possibility of error or falsehood. Instead, these algorithms learn from the probabilistic associations with a set of training data, often established by preceding algorithms. In essence, as Amoore (2020, p. 136) emphasises, when neural network algorithms in border control, law enforcement, or immigration management arrive at a decision, it is a culmination of an output signal contingent upon gradients and weighted probabilities.

As with any AI technology, the comprehension of algorithmic face recognition extends beyond mere technical aspects, such as misrepresentation and "machine bias". Instead, it should be viewed in conjunction with the larger social and political transformations within which it is conceived and implemented (Aradau & Blanke, 2022, pp. 161–181; Bueno, 2020), in light of new forms of algorithmic and statistical regimes of power.

A Brief History of Facial Recognition Technologies

According to some available sources on the history of facial recognition (Adjabi et al., 2020; de Leeuw & Bergstra, 2007; Gates, 2011), this technology began in 1964 in the US, when Woody Bledsoe, one of the founders of AI, together with his colleagues Helen Walf and Charles Bisson, in Palo Alto (California), researched the programming of computers to recognise human faces. They introduced a semi-automatic method of facial recognition that required operators to input twenty computer measures, such as mouth size or eye dimensions (Bledsoe, 1964; Bledsoe & Chan, 1965).

Computer scientists in the US continued to develop facial recognition methods during the subsequent decades, part of them working on broader research programmes in computer vision and robotics, which were themselves branches of artificial intelligence research. However, the development of facial recognition technologies only gained significant traction when the military potential of facial recognition technologies came into consideration. In 1998, the Defence Advanced Research Projects Agency (DARPA), a research and development agency within the US Department of Defense responsible for advancing emerging technologies for military applications, launched the Face Recognition Technology (FERET) programme. This initiative aimed to catalyse progress in both industry and academia. Subsequently, investment in research for military purposes was reinvigorated, with face recognition technologies gaining momentum after 9/11 (Ellerbrok, 2011, p. 529; Introna & Wood, 2004), spurred by heightened security concerns. In 2006, the Face Recognition Grand Challenge (FRGC) competition was launched with the aim of promoting the development of facial recognition technologies to support existing initiatives in this area (Adjabi et al., 2020).

As a result of growing national security concerns, CCTV systems with facial recognition capabilities have become standard in many countries around the world in locations such as airports, border crossings, stadiums and public spaces, with the aim of identifying individuals of interest, whether they are on terrorist watch lists or in national criminal databases (Ellerbrok, 2011; Kloppenburg & van der Ploeg, 2018; Magnet, 2011; Sánchez-Monedero & Dencik, 2022; Smith, 2021). Facial recognition technologies have also become a central component of biometric databases listing individuals applying for citizenship, asylum, and refugee status in several countries, such as the US and the UK (Lyon, 2010; Magnet, 2011).

Computer scientists and militaries were not the only ones interested in facial recognition technologies. By the 1990s, new companies emerged in the US to commercialize the technology, targeting markets such as institutions with proprietary computer networks (e.g., the finance industry and other business sectors) and large-scale identification systems (e.g., passport agencies, law enforcement, and penal systems) (Gates, 2011, p. 27). The objective of functioning facial recognition systems was to enhance existing identification infrastructures by replicating face-to-face recognition in networked environments. Consequently, these systems were designed to provide more accurate, effective, and ubiquitous identification systems that operate automatically, in real time, and at a distance, thereby improving institutional and administrative identification, social classification, and control (Gates, 2011, pp. 27–28).

The 2010s have seen a notable shift with the emergence of consumer-oriented facial recognition systems, which have extended the use of facial recognition beyond traditional security and surveillance contexts to find utility in everyday consumer devices such as home computers and photo-sharing platforms designed to enhance personal convenience and entertainment. In critically reflecting on the seemingly benign dimension of facial recognition technologies, it's important to consider the broader implications beyond their playful and user-friendly veneer (Ellerbrok, 2011). The author highlights how the role of "play" has historically been used to introduce controversial technologies, with the aim of familiarising individuals with high-tech security systems in non-threatening environments. This strategy is intended to reduce resistance to the introduction of such technologies, leading to their normalisation (Ellerbrok, 2011, p. 541). From Ellerbrok's perspective, engaging with these critical questions offers valuable insights into the broader social and ethical implications of the widespread adoption of facial recognition technologies.

The case of DeepFace, developed by a Facebook research team, illustrates the blurred lines between the playful, user-friendly aspects of facial recognition technologies and their potentially controversial implications. DeepFace detects human faces in digital images using an advanced neural network trained on millions of photos uploaded by Facebook users. Launched in 2015, this facial recognition feature aimed to help users identify people in their photos by automatically suggesting tags, simplifying the process of tagging friends. However, this innovation quickly sparked controversy. Critics argued that it could infringe on privacy rights and lead to unintended consequences, prompting a closer examination of the balance between technological convenience and personal privacy. As a result of growing societal concerns, in 2021 Meta, one of the largest US-based multinational information technologies, which operates several multimedia platforms, such as Facebook, Instagram, and WhatsApp, among other products and services, announced its plan to shut down Facebook facial recognition system, deleting the face scan data of more than one billion users (Metz, 2021).

While this overview of history of facial recognition technologies' development focuses primarily on the early development of these technologies in the US, it is important to consider the progress made in China. China has emerged as a global leader in facial recognition technologies and is widely recognised for its extensive and sophisticated use of these systems. Western media often highlight China's widespread use of facial recognition for mass surveillance and its role in the oppression of minorities, underscoring the broader implications and ethical concerns associated with this technology (Aradau & Blanke, 2022, p. 177; Brown et al., 2021). Understanding the history and applications of facial recognition in China provides a broader perspective on the global development and impact of these technologies.

In China the early developments of facial recognition began in 1989, when the Ministry of Public Security commissioned Professor Su Guangda of Tsinghua University to develop a "Computer Portrait Combination System". After years of research, Su Guangda's system was nationally tested in 2005, and by the 2008 Beijing Olympics, the first facial database with over ten million faces was completed.

In 2014, Tang Xiao, currently an AI industry leader in China, who earned his PhD in MIT, led a team at the Chinese University of Hong Kong to achieve a facial recognition accuracy rate that surpassed human performance. In October 2014, Tang founded SenseTime, which became the world's most valuable AI company in the late 2010s. Technological breakthroughs in the mid-2010s enabled the mass use of facial recognition for government surveillance. This period saw significant private investment in AI startups in China, the development of data collection infrastructure, widespread public and commercial use, and state deployment to monitor and police ethnic minorities (Brown et al., 2021).

In recent years, the COVID-19 pandemic has brought facial recognition technologies further into the spotlight, leading to increased deployment in many regions of the world, with China leading the way. In fact, in 2020 and 2021, many countries have spurred an increase in facial recognition applications around the world for population monitoring and epidemic surveillance. Several companies have developed facial recognition algorithms to detect compliance with mask requirements, temperature scanning, or quarantine orders (Kostka et al., 2021; Smith & Miller, 2021) to aid in the management of public health measures (Fig. 2.1).

In summary, over the past two decades, facial recognition systems have expanded beyond military, law enforcement, and security applications. They have rapidly moved into the commercial and leisure sectors and have become integrated into many aspects of daily life. Today,

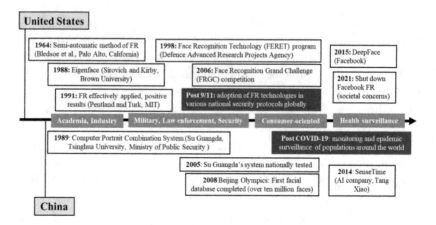

Fig. 2.1 A brief history of Facial Recognition (FR) technologies

the advancement of facial recognition technologies has spurred significant investment in a variety of sectors, including border control, retail, mobile technology, banking, and finance, among others. However, the widespread use of facial recognition technologies has been accompanied by social and ethical controversies and uncertainties. In the following sections, we will discuss how controversies and uncertainties are evolving as these technologies become more normalised and ubiquitous.

Ethical Quandaries of Facial Recognition Technologies

While developers of facial recognition technologies view this field as promising, offering numerous benefits such as enhancing safety and security, preventing crimes, locating missing persons, improving medical treatment, and streamlining shopping experiences, among many other applications, not everyone shares this techno-optimistic perspective. Advocacy groups, activists, privacy experts, scholars, and others have publicly voiced their concerns about the detrimental effects of facial recognition technologies on societies. Critics argue that these technologies normalize surveillance and erode privacy (Introna & Wood,

2004; West, 2019), exhibit significant flaws and inaccuracies (Aradau & Blanke, 2022; Cabitza et al., 2022), exacerbate discrimination, and disproportionately impact particular groups such as people of colour, women, transgender and gender non-conforming individuals, children, and people with disabilities (Buolamwini, 2023; Galligan et al., 2020; Søraa, 2023). Given the dystopian narratives associated with facial recognition technologies, there is much debate within governments, policy think tanks, and the media about whether and how these technologies should be used, and whether tighter regulation or even bans are the appropriate response. As Kostka et al. (2021) note:

> The promised benefits of FRT [facial recognition technologies] applications come with trade-offs regarding digital mass surveillance, discrimination, privacy intrusion, as well as infringement on human rights. (…) As a multi-purpose tool, the same underlying technology thus helps people and save lives and can also horrify individuals and pose ethical conundrums. (Kostka et al., 2021, p. 672)

From the perspective of facial recognition developers, the controversies and uncertainties raised so far are the result of a mere lack of technical accuracy, and therefore these errors can and must be solved by using better training datasets for machine learning (Wen & Holweg, 2023). As Bueno (2020) observes, from the developers' point of view, a properly trained algorithm accompanied by adequate technical solutions is presented as "less racist than human interactions", thus erasing historical systemic racism and social logics of power and discrimination. While facial recognition developers often attribute controversies and uncertainties to technical inaccuracies, critics argue that this perspective overlooks deeper issues of systemic racism and power dynamics. Aradau and Blanke (2022) explore the notion of "accountability by error" in the global discourse on facial recognition, suggesting that the narrative of facial recognition technologies' errors being correctable may inadvertently normalise widespread surveillance practices and obscure underlying power imbalances. Another issue relates to security concerns about the potential malicious use of facial recognition technologies (Smith & Miller, 2022). Fraud and identity theft, including spoofing and hacking, pose significant risks to facial recognition technologies. In addition, deepfake facial recognition technologies exacerbate these risks (Gupta et al., 2023), using AI to create convincing fake videos and images that can spread misinformation, defame individuals, and manipulate public opinion. Furthermore, on a personal level, deepfakes facilitate harassment, blackmail, and reputational damage, as seen in cases of targeted deepfake pornography, which particularly affects women and causes severe psychological and emotional harm.

To delve into the complexity of the contrasting perspectives on facial recognition technologies in public discourse, we provide a concise summary in Table 2.1. This table juxtaposes the views of two broadly defined social groups, without addressing their internal diversity: the developers of facial recognition technologies and their critics. Our systematization of these varied perspectives is derived from a review of the academic literature in the field of computer vision (Hupont et al., 2022; Kaur et al., 2020), which outlines the main applications of facial recognition technologies, and the critiques of these technologies by academics in the field of social sciences (Aradau & Blanke, 2022; Dauvergne, 2022).

In the upcoming section on the controversies surrounding facial recognition technologies, we will use real-world examples to carefully examine the myriad of concerns and uncertainties, ranging from privacy and civil liberties violations to bias and discrimination, potential misuse and abuse, and errors and apprehensions about accuracy and reliability.

Controversies of Facial Recognition Technologies

Facial recognition technologies have rapidly permeated various facets of modern society, from enhancing security measures to personalizing user experiences. However, this proliferation is not without significant concerns and uncertainties. Table 2.2 delineates the primary issues associated with these technologies, highlighting the critical areas of mass surveillance, the oppression and discrimination of minorities, biases and errors, and public and private partnerships.

Table 2.1 Diverse views on primary applications of facial recognition technologies

Applications	Views of developers	Views of critics
Biometric identification	Leveraged in security systems, access control, and identity verification processes	Concerns about privacy invasion, potential for misuse, and risks of surveillance, especially in relation to marginalized communities
Biometric categorization	Enables demographic analysis, customer segmentation, and targeted advertising	Raises concerns about discrimination, and the reinforcement of societal biases, particularly in advertising and consumer profiling
Emotions recognition	Used for analysis of facial expressions to discern emotions, benefiting, among others, market research and healthcare	Raises concerns regarding privacy, consent, and the potential for manipulation or exploitation, particularly in sensitive contexts like mental health care. Critics claim it is pseudo-science
Security and surveillance	Facilitates real-time identification within security and surveillance systems for public safety	Criticized for its potential for abuse, infringement on civil liberties, and the exacerbation of existing power imbalances, particularly with regard to state surveillance
Personalization and customer experience	Enhances user experiences by customizing interactions based on individual preferences	Raises concerns about privacy infringement, consent issues, and the creation of filter bubbles, particularly in the context of data collection and targeted advertising

(continued)

Table 2.1 (continued)

Applications	Views of developers	Views of critics
Attendance tracking and time management	Automates attendance tracking and time management, streamlining administrative processes in workplaces and educational settings	Raises concerns about employee and student privacy, surveillance in the workplace or educational institutions, and the potential for misuse or abuse of attendance data
Healthcare applications	Facilitates patient identification, condition monitoring, and disease diagnosis in healthcare	Concerns about data security, patient privacy, and the reliability and accuracy of diagnosis, especially in the context of sensitive medical information and vulnerable populations

Mass Surveillance with Facial Recognition Technologies

Facial recognition technologies have often been framed in public debates as tools for mass surveillance, used by governments, law enforcement agencies, and corporations alike. These controversies present scenarios ranging from the extensive surveillance practices of authoritarian regimes such as China and Russia to widespread surveillance in liberal democratic societies. Some examples of areas where facial recognition technologies are expanding and raising concerns about privacy and civil rights issues include:

Smart Cities: In many cities around the world, facial recognition technologies are being integrated into various aspects of urban infrastructure as part of smart city initiatives. For example, it may be used to monitor traffic flow, identify wanted persons, or enhance security in public areas.

Law enforcement: Law enforcement agencies in several countries are using facial recognition technology to identify suspects or persons of interest in crowds or public spaces. For example, police departments may

Table 2.2 Controversies surrounding facial recognition technologies

Controversy category	Main elements
Mass surveillance	Invasion of privacy Surveillance of protests and crowds Potential for government overreach Lack of consent from individuals Proliferation of surveillance infrastructure
Oppression and discrimination of minorities	Disproportionate targeting of marginalized communities Amplification of existing biases in policing and law enforcement Potential for exacerbating social inequalities and reinforcing stereotypes and discrimination Impact on civil liberties and human rights
Bias and errors	Racial and gender biases in algorithmic decision-making Inaccuracies, errors, frauds, and fakes Lack of transparency in algorithms and decision-making processes
Public and private partnerships	Lack of clear regulation Insufficient transparency Private companies cooperating with law enforcement agencies without clear oversight

deploy facial recognition systems at events, protests, or transport hubs to scan faces against databases of known individuals.

Airport security: Many airports have implemented facial recognition technologies as part of their security measures. These technologies are used for tasks such as identity verification at check-in, security screening, and boarding. Some airports also use facial recognition for surveillance purposes to monitor passenger movements throughout the terminal.

Retail and marketing: retailers and marketing companies are using facial recognition technologies to analyse customer behaviour and preferences. For example, stores can use facial recognition cameras to track how customers navigate the store and what products they are interested in. This data can then be used to optimise store layouts and marketing strategies.

A highly controversial use of facial recognition technologies, which can be applied in the above sectors as well as in many others—for example, in healthcare, schools, and law enforcement—is the use of these technologies to **assess emotional states, infer behavioural tendencies, and predict personality traits.** Some notable examples include emotional state assessment in healthcare and therapy, where therapists can use facial recognition to monitor patients' emotional states in real time during sessions to help understand patient reactions and tailor therapeutic interventions. A number of mental health apps use facial recognition to assess users' emotional states and provide appropriate responses or exercises to manage stress, anxiety, and depression. In education, facial recognition technologies could be used to monitor students' emotional engagement to help teachers adjust their teaching methods and ensure that students are attentive and interested. Inference of behavioural tendencies could be used by security and law enforcement agencies to detect signs of stress or anxiety that may indicate suspicious behaviour in public places such as airports or large events. At large gatherings, facial recognition could also help identify people who are behaving abnormally, allowing security personnel to intervene proactively. However, due to the highly problematic and biased nature of such applications, it is not yet known how widespread they will become. Critics argue that because emotions exist only as subjective experiences, the underlying assumptions of automated emotion recognition have major conceptual problems and are unreliable. As noted by David Black (2023):

> Emotion recognition attempts to shift surveillance from scrutiny of the bodily exterior to scrutiny of internal subjective states, and from the recording of behaviour to its prediction. The fact that this promise rests more on fantasy and pseudoscience than reality highlights the degree to. which any attempt to automate our understanding of faces will always be fundamentally compromised by both the complexity of human perception and the face's refusal to provide any stable unit of study or measurement. (Black, 2023, p. 1439)

In sum, societies characterized by the widespread deployment of facial recognition technologies anticipate the normalization of forms of mass-customized biopolitics (Andrejevic et al., 2024), and of targeted governance: the ability to operate on the population and the individual simultaneously through automated forms of passive identification and machinic recognition and analysis.

The case of China: A Stark Illustration of Facial Recognition Surveillance

China's emergence as a leader in facial recognition technologies has heightened concerns about dystopian mass surveillance, intensifying debates in Western countries about the balance between freedom and repression, and the intersection of surveillance and democracy. These concerns are greatly amplified by media narratives that often portray China's use of AI for domestic surveillance as a stark example of state control, serving as a poignant contrast to the envisioned ethical applications of AI and the democratic ideals presumably embodied by the US and the European Union (EU) (Aradau & Blanke, 2022, pp. 176–177).

The Chinese government aims to establish a comprehensive network of facial recognition cameras, boasting continuous operation and full control. With nearly two hundred million public surveillance cameras already in place, China's surveillance infrastructure is extensive, capable of identifying individuals within crowds (Aradau & Blanke, 2022, p. 177). Notably, policing applications of facial recognition in China often receive positive coverage in the Chinese media. As Peter Dauvergne (2022, p. 72) points out, the Chinese media frequently highlight the effectiveness of AI-powered facial recognition technology in locating missing children. One example is Wang Yiqing's article, "Face-Recognition Technology Reunited More Than 10,000 Lost People" (The Straits Times, January 21, 2020), which discusses how Baidu's technology, used by China's Ministry of Civil Affairs and local rescue centres, helps locate missing children by matching decades-old photographs with recent images of adults. This method has been praised for enabling the reunification of thousands of missing children with their families (ibid., p. 72).

As noted above, China's adoption of facial recognition technologies, like many other digital advances, is often portrayed as a grim embodiment of pervasive surveillance, supposedly in stark contrast to Western advances in AI and facial recognition. However, as Aradau and Blanke (2022, p. 161) highlight, global concerns about facial recognition tend to reproduce geopolitical divisions and forms of othering, with China often assumed to be advancing its facial recognition capabilities unimpeded. The authors further note that as China has emerged as a frontrunner in facial recognition technologies, depictions of dystopian surveillance have intensified in the media and Western discourse (Aradau & Blanke, 2022, p. 177). In light of this, the question arises as to whether our understanding of China's use of facial recognition technologies is primarily shaped by narratives—whether media or political—that seek to draw a clear distinction between dystopian applications in China and those in Western nations. Conversely, the widespread integration of facial recognition applications—in China facilitating numerous everyday tasks such as payments, access control, and airport procedures, echoing similar trends in Western nations and beyond—raises concerns about the increasingly blurred distinctions between mass surveillance using facial recognition technologies in China and the diverse realities of the "West". This calls for a re-evaluation of the controversies surrounding facial recognition technologies, whether they operate within authoritarian or democratic frameworks.

This re-evaluation is further complicated by the reality that surveillance and resistance often coexist. In China, despite the extensive use of facial recognition technologies for surveillance, there have been notable instances of resistance against these systems. This resistance does not negate the existence of pervasive surveillance but instead underscores the tension between state control and individual pushback, highlighting how surveillance practices are contested even in highly controlled environments (Aradau & Blanke, 2022, pp. 178–179). Therefore, it is essential to recognize that the presence of resistance to facial recognition technologies in China signals an ongoing struggle within the framework of state surveillance, not a denial of its existence.

Guo Bing, a law professor, initiated China's first privacy lawsuit against the Hangzhou Safari Park for using facial recognition to replace fingerprint scans without explicit consent. This lawsuit raised concerns among Chinese citizens about the widespread use of facial recognition. Despite court rulings in Guo's favour, ordering the park to delete his data and refund his fees, he continued to appeal, challenging the park's authority to collect facial data (Chen & Wang, 2023).

Similar cases, such as law professor Lao Dongyan's rejection of facial recognition in residential areas, have also attracted attention in China. Meanwhile, the enactment of China's new Personal Information Protection Law (PIPL), which came into effect in November 2021, aims to strengthen the protection of personal data. The PIPL applies to organisations and individuals that collect, use, store, transfer, or disclose the personal information of individuals located in China. The PIPL specifically covers the use of facial recognition in public spaces, including that it can only be used for public security purposes, unless each individual has given individual consent. However, as noted by Brown et al. (2021), different legal standards apply to public and private uses of facial recognition in China. State security organs can collect any information they deem necessary in public or for crime investigation, with relatively weak privacy laws protecting citizens from police surveillance compared to the US. While police must keep personal information confidential, there are few restrictions on how they collect or use it internally. The public debate on facial recognition focuses on its use by private organisations, which is often accepted when it is in line with state security priorities.

As noted by Dauvergne (2022, p. 75), despite some victories for privacy and consumer rights advocates, China is steadily normalizing facial recognition technologies. It is becoming an unquestioned, even expected and demanded, part of politics, the economy, and everyday life.

Global Expansion of Facial Recognition Technologies

Other countries are following in China's footsteps in expanding facial recognition technologies. Many have accelerated their pace in recent years, particularly since the COVID-19 pandemic. This global trend reflects a growing acceptance and integration of these technologies into various sectors, raising privacy and civil rights concerns similar to those in China.

A notable example of the expansion of facial recognition technologies during the COVID-19 pandemic, as well as concerns about their continued use, is the case of Australia (Andrejevic et al., 2020). Since 2019, the increased use of facial recognition in Australia has faced significant backlash due to its failure to protect the privacy of citizens. Facial recognition was widely used in public spaces during the COVID-19 pandemic without proper legal oversight, and this trend has continued since. The Australian Federal Police, as well as various state and territory forces, use facial recognition systems extensively to identify persons of interest in public spaces. In 2023, facial recognition returned to the parliamentary agenda when the Attorney-General introduced the Identity Verification Services Bill (IVS). The government announced its commitment to introducing legislation to protect the personal information of Australians. Some experts have noted that China's Personal Information Protection Law 2021 provides significantly stronger privacy protections than Australia's (Field, 2023).

In 2024, law enforcement and government agencies in Australia dealt with a major breach of personal data related to a facial recognition system used in bars and clubs. The company involved was Outabox, an Australian company with offices in the US and the Philippines. In response to the Covid-19 crisis, Outabox introduced a facial recognition terminal to screen visitors and monitor their body temperature. These terminals were also used to identify people participating in a self-exclusion programme for problem gamblers. Recently, a website called "Have I Been Outaboxed" surfaced, allegedly created by former Outabox developers in the Philippines. Visitors to the site can enter their names to find out if their details have been included in an alleged Outabox database, which is said to lack strict internal safeguards and is shared via an unsecured spreadsheet containing over 1 million records (Pearson, 2024).

Prominent cases of mass surveillance involving partnerships between governments and private companies in liberal democracies have raised additional concerns. One notable example is Amazon's Rekognition software, which faced significant backlash for being sold to law enforcement agencies without clear oversight, raising serious privacy and civil liberties concerns. In the UK, police and private companies like Tesco, Budgens, and Sainsbury have used facial recognition technologies in public spaces such as shopping malls and stadiums, facing legal and public opposition over privacy concerns. Startups like Yoti and Facewatch also use facial recognition technologies for age verification in cinemas and sharing CCTV images with law enforcement. The UK Information Commissioner's Office ordered public service providers Serco Leisure, Serco Jersey, and related leisure trusts to stop using facial recognition technologies and fingerprint scanning for employee attendance in 2024 due to privacy issues.

In addition, Clearview AI, a US-based company, has been at the centre of widespread controversy and scrutiny for its practices around facial recognition technologies and the problematic relationships between private companies and public entities. Clearview AI has come under fire for its massive data collection methods, which involve scraping billions of images from social media platforms and other online sources without users' consent (Milmo, 2022). Clearview AI's practices have raised deep concerns about privacy violations and the unchecked proliferation of facial recognition capabilities. Furthermore, Clearview AI's technology has been used by law enforcement agencies in the US and around the world, raising significant ethical and legal questions about the implications of mass surveillance and the erosion of civil liberties. It has also been reported that in addition to providing facial recognition software to law enforcement agencies, Clearview AI may also provide services to private companies in the US (Smith & Miller, 2022).

Critics argue that Clearview AI's practices pose serious risks to individuals' rights to privacy and anonymity, highlighting the urgent need for robust regulation and oversight of facial recognition technologies to guard against abuse and protect fundamental freedoms (Dauvergne, 2022, pp. 59–62). This imperative was further underscored by the release of a "Consensus Study Report" in January 2024 by the Committee on Facial Recognition of the National Academies of Sciences, Engineering, and Medicine. The report, sponsored by the US Department of

Homeland Security and the Federal Bureau of Investigation, identified significant concerns about certain uses of facial recognition technologies that warrant immediate government attention (National Academies, 2024). The report's recommendations for federal legislation, executive orders, and multi-stakeholder engagement signal a concerted effort to address the challenges posed by facial recognition technologies and to ensure their responsible development and deployment in line with societal values and priorities. However, this regulatory attempt could be seen as driven by a desire to promote public trust and acceptance of AI developments, working to stabilise or end controversies without changing the underlying operations of facial recognition development (Machado et al., 2023).

Private–public partnerships are not limited to the US and the UK. In Brazil, the provision of biometric surveillance systems in city administrations is often delegated to private companies that provide facial recognition technologies to law enforcement agencies. These partnerships lack transparency and rely on the legitimacy of the public administration to provide services (Ramiro & Cruz, 2023). Similarly, in South Africa, the privatisation of public surveillance has sparked debates about its impact on democratic freedoms. Civil liberties activists warn that the proliferation of advanced surveillance technologies is creating a form of digital apartheid where privacy and security are unequally distributed (Hao & Swart, 2022).

Opposition to Facial Recognition Technologies

More than 20 municipalities in the US, from San Francisco to Boston, have banned or severely restricted the use of facial recognition technologies in law enforcement and public services. Portland, Oregon, has taken an even more far-reaching approach by banning private companies from deploying the technology in public areas (Dauvergne, 2022, p.11). Despite these bans on the use of facial recognition by government agencies in US cities, the technology continues to spread to police departments, federal agencies, schools, and private businesses across the country (Dauvergne, 2022, p.42).

The simultaneous calls to ban facial recognition technologies and their widespread use illustrate the complex relationship between democracy and surveillance, and highlight the increasingly blurred line between freedom and repression (Aradau & Blanke, 2022). A notable example of these blurred lines is the controversy surrounding the use of facial recognition to monitor public protests. A famous case is the global attention given to the Black Lives Matter movement, which highlighted issues of police brutality and racially motivated violence against black people. The protests that erupted in the US in 2020 following the killing of George Floyd by Minneapolis police officers intensified this scrutiny. Tech giants involved in the development of facial recognition technologies were criticised for working with law enforcement to identify protesters. This intersection of events underscores broader concerns about the ethical implications of facial recognition technologies, prompting debates about civil liberties and democratic values in an increasingly digitalised world. In response to public outrage, Amazon declared a moratorium on police use, IBM halted development of facial recognition systems, and Meta announced plans to discontinue Facebook's facial recognition system and delete the facial scan data of over one billion users. However, these reactions from big tech companies are seen by critical commentators as strategies to mitigate or prevent reputation loss (Wen & Holweg, 2023).

In Europe, a coalition of over 110 civil society organisations across 25 EU countries, united as the European Digital Rights Network, started in November 2020 a campaign called "Reclaim Your Face", calling on authorities across the EU to impose a comprehensive and permanent ban on public facial recognition technologies, with targeted campaigns in countries including the Czech Republic, France, Italy, Germany, Greece, the Netherlands, Serbia, and Slovenia. While calls for a ban on facial recognition technology are growing across Europe, its use continues to soar in the region and is widely seen as essential for law enforcement and border security. A telling example of the nuanced interplay between resistance to facial recognition and its implementation occurred in June 2023, when the European Parliament made a historic decision to support a blanket ban on live facial recognition in public spaces. However, despite the expectations of activists and some parliamentarians advocating for a ban or strict regulation, the EU's AI law was amended at the last minute in January 2024 to allow law enforcement to use facial recognition technology on recorded video footage without judicial approval.

In the UK, a coalition of NGOs, including prominent organisations such as Amnesty International, Big Brother Watch, Liberty, the Institute of Race Relations, the Open Rights Group, and Privacy International, have taken legal action to challenge the use of live facial recognition by law enforcement and security agencies, and are actively pressuring the UK government to ban the use of live facial recognition technology (Dauvergne, 2022, p. 53). This backlash against live facial recognition has been fuelled by controversies surrounding its use in the UK. Trials conducted by the Metropolitan Police Service (MPS) between 2016 and 2019 involved the use of fixed cameras to scan the faces of passers-by and compare the captured images with a watchlist of wanted individuals (Bradford et al., 2020), sparking protests and criticism from civil society organisations.

Recent incidents in Russia challenging facial recognition technologies include the alleged use of facial recognition to identify and detain protesters ahead of demonstrations following the arrest of opposition leader Alexei Navalny in April 2021. Organisations such as the Russian digital rights NGO Roskomsvoboda have campaigned for a moratorium on facial recognition technology for mass surveillance. Roskomsvoboda argued that the proliferation of this technology in Russia poses a threat to civil society by stifling dissent and protest, collecting data without explicit consent, and contributing to the violation of citizens' rights by both security forces and corporations (Dauvergne, 2022, pp. 53–54).

In some Latin American countries, activism against facial recognition is on the rise (AI Sur, 2021; Caiero, 2022). In Brazil, the #TakeMyFace-OffYourGaze campaign (#TireMeuRostoDaSuaMira) was launched in 2022 to draw public attention to the potential risks of using facial recognition, bringing together around 60 NGOs advocating for a ban on facial recognition technologies. The campaign points to evidence of abusive and non-transparent use of facial recognition systems in Brazil, noting that all Brazilian states currently use facial recognition in public security to identify people suspected of crimes or with outstanding warrants (Tiremeurostodasuamira, 2022). In Argentina, several civil society organisations complained about the abusive use of facial recognition systems by the government. In 2023, the civil society organisation Fundación Vía Libre, which aims to defend human rights by acting as a watchdog on public policies in the field of ICTs, stated that Argentina was using facial recognition systems to prevent demonstrations (https://www.vialibre. org.ar/nosotros/). The Mexican civil society organisation #NoNosVean-LaCara (Red en Defensa de los Derechos Digitales, 2020) has proposed a ban on facial recognition technologies (Ramiro & Cruz, 2023, p. 12).

Minority Oppression and Discrimination

Another deeply controversial aspect of facial recognition technologies is their association with the oppression and discrimination of minorities. A common narrative around the risks of facial recognition technologies revolves around the use of advanced Chinese facial recognition systems to monitor the Uyghur minority, which has been widely denounced as "the first known example of a government deliberately using artificial intelligence for racial profiling" (Mozur, 2019). Chinese security agencies have used automated facial recognition to track and profile residents of Xinjiang, a region in northwest China inhabited by Muslim Uyghurs. In other parts of China, facial recognition technology is being used to track people who resemble Uyghurs. CloudWalk, a Chinese startup, explains how this software works: "If a Uyghur originally lives in a neighbourhood and six Uyghurs are detected within 20 days, it immediately triggers an alarm" to the police for possible prosecution and detention (Dauvergne, 2022, p. 2). Social media platforms use AI-driven facial recognition to flag content containing "sensitive individuals", such as those who appear to be Uyghurs, in order to comply with government regulation.

In February 2022, Amnesty International reported that Israel's system of oppression and domination of Palestinians amounts to apartheid. According to the international human rights NGO, Israeli authorities use facial recognition and other biometric technologies to monitor and restrict Palestinians, impeding their freedom of movement and access to basic rights such as work, education, and healthcare (Amnesty International, 2022). In Hebron and East Jerusalem, where Israeli settlements exist, pervasive surveillance is used to pressure Palestinians to leave strategic areas. This surveillance is also used to identify and target peaceful demonstrators, violating the right to freedom of expression and assembly. The Israeli authorities further tighten control by establishing extensive CCTV networks, watchtowers, and checkpoints, creating Palestinian-specific databases, and using facial recognition software for mass and discriminatory surveillance (Brooks & Griffiths, 2024).

However, the tracking and surveillance of ethnic and racial minorities might be disguised as measures to ensure public safety and security, rather than being openly acknowledged as monitoring minorities. For instance, a frequently cited example in media outlets and human rights-focused non-governmental organizations, particularly those based in the US, is the use of facial recognition software in New York City (Jackson, 2024). This technology is employed by the New York City Police Department (NYPD), which describes its capabilities as follows:

> Since 2011, the NYPD has successfully used facial recognition technology to investigate criminal activity and increase public safety. The NYPD uses facial recognition to aid in the identification of suspects whose images have been recorded on-camera at robberies, burglaries, assaults, shootings, and other serious crimes. The NYPD also uses facial recognition to aid in the identification of persons unable to identify themselves (e.g., persons experiencing memory loss or unidentified deceased persons). (NYPD 2023, p. 4)

From the perspective of civil society organisations, the use of facial recognition technology in New York is often described as a threat to the right to protest and as exacerbating racially discriminatory policing practices by putting black and minority communities at risk of misidentification and wrongful arrest, with very high reported inaccuracies. In addition, property owners and managers overseeing rental properties in New York are using facial recognition technology to monitor tenants or individuals in their properties, particularly targeting black and brown communities (Amnesty International, https://banthescan.amnesty.org/nyc/).

Bias and Errors

The discourse surrounding facial recognition technologies has become increasingly prominent due to their potential to perpetuate discrimination and produce inaccurate results. Studies reveal that these systems often exhibit significant flaws and inaccuracies (Cabitza et al., 2022), disproportionately affecting people of colour as these technologies are structurally biased towards whiteness (Pugliese, 2010). In addition, women, trans and gender non-conforming people, children, and people with disabilities are also negatively impacted (Galligan et al., 2020). The inherent biases in facial recognition technologies can lead to discriminatory practices, such as wrongful arrests or unequal treatment in various scenarios, including hiring processes and security screening. Concerns about these biases are particularly evident in the US government's new mobile app, CBP One, which is used by migrants to apply for asylum at the US-Mexico border. Reports suggest that the app disproportionately prevents many Black people from submitting their claims due to the technology's inherent biases. Non-profit organisations advocating for Black asylum seekers point out that the app fails to register people with darker skin tones, effectively denying them the opportunity to apply for entry into the US. This issue predominantly affects individuals from Haiti and African countries, highlighting the algorithmic biases present in the technology (del Bosque, 2023).

Computer scientist Joy Buolamwini, founder of the Algorithmic Justice League in the US, has been instrumental in exposing the racial and gender bias inherent in facial recognition, highlighting the disturbing reality that bias can exist within these systems without proper oversight or scrutiny (Buolamwini, 2023). This disturbing trend has even caught the attention of government scientists, who now acknowledge the flawed and biased nature of this surveillance technology. Jay Stanley, a policy analyst at the American Civil Liberties Union (ACLU), highlights the far-reaching consequences of such biases. A single false match, he notes, can lead to a cascade of negative outcomes, including missed flights, prolonged interrogations, placement on watch lists, tense police encounters, wrongful arrests, or worse (Dauvergne, 2022, p. 6).

This issue of bias and error has received considerable attention in both the US and Europe. In the UK, for example, a report by civil liberties organisation Big Brother Watch (2018) revealed alarming statistics: facial recognition systems used by the police misidentified individuals nine out of ten times, highlighting the systemic biases inherent in these technologies. In addition, research by Leslie (2020) and Buolamwini and Gebru (2018) highlighted how algorithms can inadvertently encode race and class biases and present them as impartial truths. This phenomenon is not confined to the UK, but resonates across borders, as evidenced by discussions within the EU. A study commissioned by the LIBE Committee highlighted the challenges associated with the integration of digital facial data into police exchanges in the EU, citing concerns about privacy, data security and the potential for discrimination against certain demographic groups (Vavoula, 2020).

In 2022, the European Data Protection Board (EDPB) raised concerns about the use of facial recognition technology in law enforcement, stressing the need to carefully consider challenges related to reliability, efficiency, and the overall quality and accuracy of data sources. The EDPB's opinion underlines the significant risks for individuals involved in law enforcement activities. It highlights how public fears about "errors" and the inadequacy of datasets used to train algorithms have manifested themselves in controversies surrounding facial recognition (EDPB, 2023). The EDPB (2023) emphasised that facial recognition technologies, whether used for authentication or identification purposes, do not provide definitive results, but rely on probabilities to determine whether two faces or images belong to the same person. In addition, the EDPB noted that various studies have shown that such statistical results derived from algorithmic processing can be subject to bias, which can arise from factors such as the quality of the source data, the training databases or even the location of use. This underlines the complexity of ensuring the fairness and accuracy of facial recognition systems in law enforcement contexts.

Facial Recognition Around the World and Instances of Resistance

With advances in recognition algorithms and the proliferation of facial databases, the adoption of facial recognition technologies continues to grow globally, permeating different sectors and with varying levels of demand, acceptance, and resistance in different regions. However, comprehensive empirical data on their applications remains scarce, often limited to anecdotal reports and media coverage. Comparitech, a consumer advocacy company that promotes online privacy and cyber-security, educates consumers on various cybersecurity issues, and extensively tests and reviews cybersecurity products and services, undertook a comprehensive study to understand the prevalence of facial recognition technologies in each country.

The Comparitech study looked at the use of facial recognition technologies in the 100 most populous countries in the world, covering government, law enforcement, airports, schools, banks, workplaces, buses, and trains (Bischoff, 2022). North Korea was excluded from the analysis due to insufficient data clarity. The study found that almost 80% of countries have growing, widespread, or invasive government use of facial recognition technologies. Access to some form of these technologies was also common at banks or financial institutions and airports (almost 80% and 60% of countries, respectively), followed by workplaces (around 40% of countries). The use of facial recognition technologies was less common in trains or subways (30% of countries), buses (20% of countries), and schools (almost 20% of countries). In Bischoff's, 2022 analysis, each of the 99 countries received a score out of 40, with higher scores indicating minimal or less invasive use of facial recognition technology and lower scores indicating widespread and invasive use. Of the 99 countries analysed, the following countries were found to have the most widespread use: China (5); Russia (9); the United Arab Emirates (10); Chile, India, and Japan (12); Australia and Brazil (13); Argentina (16); France, Hungary, Malaysia, and the UK (17); Mexico and the US (18); Romania, Spain, and Taiwan (19); Kazakhstan, Sweden, South Africa, and Thailand (20).

The continued development and expansion of facial recognition technologies necessitates an examination of the evolving digital politics of resistance (Amoore & Hall, 2010). These resistances manifest themselves in different ways, not only as outright opposition, but also as efforts to redirect power relations. Resistance can be individual or collective, revolutionary or transformative. Table 2.3 summarises different geographies of resistance, based on our analysis of media articles, NGO reports, and the limited academic literature available.

Networks and Ethical Assemblages Around Facial Recognition Technologies

In this section, we explore how the controversies and uncertainties surrounding facial recognition technologies and the resistance they face create a complex, dynamic, and diverse landscape that opens up debates and public contestation and attempts to resolve them. Our analysis is rooted in the concept of ethical assemblages discussed in Chapter 1, which refers to the unstable and evolving collective understandings of what is considered ethical, unethical, hidden or suppressed in relation to facial recognition technologies. These ethical assemblages emerge and take shape through values, discourses, interactions, and practices within specific networks. These networks involve a variety of social actors, each situated in their own political interests and power relations (Bigo et al., 2019), and continuously reconfigure relationships between states, regulators, technology companies, civil society organisations, and citizens. Table 2.4 provides a systematization of networks and the various actors involved, and the related ethical assemblages that will be analysed in this section.

To elaborate on the dynamics summarised in Table 2.4, we draw on insights from Science and Technology Studies (STS) to analyse the power relations and interests that drive the opening and closing of controversies and uncertainties over facial recognition technologies. We do this by focusing on the networks and assemblages, with the aim of tracing how different actors negotiate and contest the meanings and implications of these technologies. This approach highlights how ethical issues are

Table 2.3 Cases of resistance to facial recognition technologies

Countries	Controversies	Social actors
Argentina[a]	– Rapid uptake of facial recognition for police surveillance in public spaces: football stadiums (2018); identify individuals with criminal records or arrest warrants; recognize fugitives and missing persons at public transport terminals in Buenos Aires (2019, declared unconstitutional in 2022) – Infringements to privacy and lack of consent – Deterrence of demonstrations	– Civil society organizations (e.g., Fundación Vía Libre)
Australia[b]	– Widespread use of facial recognition since 2019 without proper legal oversight – Infringements to privacy, including breach of personal data tied to a facial recognition used in bars and clubs	– Government agencies introduce legislation aimed at protecting the personal information of Australians (2023)
Brazil[c]	– Rising trend in the usage of facial recognition since 2019 – Abusive and non-transparent use of facial recognition: identify people suspected of committing crimes or who have outstanding arrest warrants; public–private partnerships	– Civil society organizations and human rights NGOs (e.g., Igarapé): campaign "#TakeMyFaceOffYourGaze" (2022); "right to the city"

(continued)

Table 2.3 (continued)

Countries	Controversies	Social actors
China[d]	– Infringements to privacy and lack of consent (Safari Park, residential settings) – Different legal standards apply to public and private uses of facial recognition – Relatively weak privacy laws protecting citizens from police scrutiny	– Law professors – Privacy and consumer rights advocates – Regulators and policymakers: personal Information Protection Law (2021)
France[e]	– Use of facial recognition in schools: inability of students to freely consent due to the authority relationship with the school administration – AI video surveillance in public spaces during the 2024 Olympic Games	– A court in Marseille (2020) overturned regional authorities' decision to install facial recognition in two high schools
Kazakhstan[f]	– Enhancing surveillance infrastructure through partnerships with Chinese ICT companies: concerns regarding the transparency and control of data storage practices	– Legislative assurances regarding data sovereignty
Mexico[g]	– Involvement of various companies in the implementation of facial recognition that have been internationally questioned for their alleged involvement in human rights violations – Lack of transparency	– Civil society organizations and academia (e.g., AI Sur and #NoNosVeanLaCara) proposed a ban of facial recognition
Russia[h]	– Infringements to privacy and lack of consent – Stifle protesters and dissents – Exposure to potential future harassment	– Digital rights non-governmental organizations (e.g., Roskomsvoboda) – Activists' complaints to the European Court of Human Rights about the use of facial recognition during protests

(continued)

Table 2.3 (continued)

Countries	Controversies	Social actors
South Africa[i]	– Privatized public surveillance: lack of regulation and technical limitations – Digital apartheid: access to privacy and security is unevenly distributed	– Civil rights activists
Spain[j]	– Legality of employment of facial recognition by retailers to aid law enforcement in managing access for individuals with criminal records or under suspicion: absence of clear regulations and transparency	– Data protection authorities
Sweden[k]	– Using facial recognition to track student attendance in a school (2019) – Use of facial recognition by Police Authority: unlawful use of Clearview AI for personal data processing	– Swedish Authority for Privacy Protection (SAPP): (a) concluded that the test violated several articles of the GDPR and imposed a fine on the municipality; (b) confirmed unauthorized use of Clearview AI, and organizational measures were found insufficient to ensure compliance with the Criminal Data Act (e.g., no data protection impact assessment)
Taiwan[l]	– Robust and extensive surveillance infrastructure: privacy concerns; prevalence of blacklist Chinese-manufactured cameras; fears of potential surveillance and data sharing by China	– City officials have committed to ensuring compliance with security standards and regulatory provisions

(continued)

Table 2.3 (continued)

Countries	Controversies	Social actors
Thailand[m]	– Potential misuse of pervasive digital surveillance measures and biometric data: reinforce government's efforts to stifle civic space and protesters, and to monitor the activities of human rights defenders and political dissidents; potential threats to privacy and civil liberties	– Human rights organizations and civil society groups
United Kingdom[n]	– Collaboration between private digital service providers and public entities, especially in law enforcement – The former UK Biometrics and Surveillance Camera Commissioner, the watchdog responsible for monitoring facial recognition in the UK, joined Facewatch, a facial recognition firm (2023) – Large-scale deployment of live facial recognition by law enforcement and security agencies generating concerns about public privacy rights (for example, the King's Coronation in May 2023, where live facial recognition use was described as signalling the UK government's support for expanding its usage to deter and detect crime in public settings with large crowds)	– The UK Information Commissioner's Office instructed public service providers to cease utilizing facial recognition for monitoring employee attendance (2024) – The Information Commissioner (ICO) opinion on the use of live facial recognition in public places identified several key data protection issues (2021) – Civil society organizations and NGOs (e.g., Amnesty International, Big Brother Watch, Liberty, the Institute of Race Relations, the Open Rights Group, and Privacy International) pressed the government to ban the use of live facial recognition

(continued)

Table 2.3 (continued)

Countries	Controversies	Social actors
United States[o]	– AI-driven systems integrated with facial recognition and drone technology in border enforcement – Unbridled expansion of surveillance capabilities that extends far beyond border regions into domestic spheres – Algorithmic racial and gender bias, and misidentification – Lack of oversight or scrutiny – Privacy infringements, lack of consent, and mass surveillance (e.g., massive data collection methods enacted by Clearview AI, a US-based company, without users' consent, whose technology has been employed by law enforcement agencies and private companies)	– Government scientists acknowledged the flawed and biased nature of facial recognition – Civil liberties and privacy advocates, US lawmakers, non-profit organizations helping asylum seekers, and Algorithmic Justice League: urgent need for comprehensive regulation and oversight – Report of the National Academies of Sciences, Engineering, and Medicine, sponsored by the Department of Home and Security and the Federal Bureau of Investigation (January 2024): federal legislation, executive orders, and multi-stakeholder engagement

Sources [a] Caeiro, 2022; O.D.I.A., 2022; [b] Field, 2023; [c] Caeiro, 2022; Instituto Garapé, 2023; Ramiro & Cruz, 2023; Tiremeurostodasuamira, 2022; [d] Aradau & Blanke, 2022; Dauvergne, 2022; [e] Fouquet, 2019; Nikolic, 2020; [f] Kassenova & Duprey, 2021; Stryker, 2021; [g] AI Sur, 2021; Flores, 2021; Ramiro & Cruz, 2023; [h] Aradau & Blanke, 2022; Bischoff, 2022; Dauvergne, 2022; [i] Hao & Swart, 2022; McConvey, 2024a; [j] Martínez, 2022; [k] Eneman et al., 2022; Nash, 2023a, 2023b; Polisen, 2024; [l] Bauer, 2023; Macdonald, 2023; Strong, 2023; [m] Abhishek, 2024; ECNL, 2023; McConvey, 2024b; [n] Bradford et al., 2020; Dauvergne, 2022; Gentile, 2023; ICO 2021; [o] Buolamwini, 2023; CBP, 2024; Dauvergne, 2022; del Bosque, 2023; Milmo, 2022; National Academies, 2024; Smith & Miller, 2022; Tyler, 2022

continually constructed, contested, and redefined, making these controversies and uncertainties integral to the very fabric of technological innovation and governance of facial recognition technologies. Overall, the field of controversy studies within the social studies of science and technology has significantly broadened its scope beyond its initial focus on factual disputes primarily involving competing expert communities

Table 2.4 Networks and ethical assemblages of facial recognition technologies

Actors	Networks	Ethical assemblages
Academic scientists	–Conduct fundamental and applied research in AI and facial recognition technologies –Engage in collaborations with other academic institutions and industry partners –Participate in academic and professional organizations, influencing standards and practices in the field and adoption	–Adhere to ethical guidelines and codes of conduct set by academic and professional bodies, and advocate for transparency in AI development processes –Contribute to public discourse on the ethical implications of AI and facial recognition technologies –Engage with policymakers to inform regulations and ethical standards based on scientific evidence and research
Big tech companies	–Develop and deploy AI facial recognition technologies –Control data necessary for training algorithms –Shape market availability and adoption	–Collaborate with governments on projects –Lobby policymakers to shape favourable regulations; face regulatory tensions –Need public trust; concerns over privacy and consent lead to demands of transparency
General citizens	–Provide data, often unknowingly, used in facial recognition technologies –Influence policies through resistance and acceptance	–Public opinion affects market adoption and regulatory policies –Mobilized by NGOs to demand transparency and ethical standards

(continued)

Table 2.4 (continued)

Actors	Networks	Ethical assemblages
Governments	–Use facial recognition for law enforcement, border control, public safety, military purposes –Provide funding for research and development	–Influence public trust through surveillance practices; public opinion and activism influence policy change –Collaborate with tech companies and research institutions
Non-governmental organisations	–Advocate for ethical use and raise awareness on potential abuses of tech companies and governments –Conduct independent research and publish critical analysis	–Scrutinize activities of big tech companies and governments –Influence policies through research and advocacy –Inform and mobilize citizens for ethical practices and policies
Regulators and policymakers	–Create regulatory frameworks for ethical use of facial recognition technologies –Ensure compliance with laws and regulation	–Ensure compliance and accountability from technology companies and government use

(Jasanoff et al., 1995). In our view, controversy studies should secure a crucial and enduring role in the analysis of AI and society. The case of facial recognition technologies exemplifies this need, highlighting how controversies and uncertainties surrounding these technologies intersect with ethical, social, and political dimensions, making them essential for ongoing scientific inquiry.

First, by considering the positioning of academic researchers working on facial recognition technologies, who tend to play a hybrid role: contributing critical perspectives on ethical implications and biases and advocating for ethical standards, while some of them maintain relationships with industry and contribute to validating and supporting the development of these technologies. Second, through the techno-solutionism offered by technology companies, which includes a strategic emphasis on error and accountability. Third, through regulatory attempts involving civil society organisations and advocacy groups focused on the protection of human rights and civil liberties. Fourth, by consulting the general public to understand their concerns and expectations about facial recognition technologies, with the overarching goal of ensuring that the use of facial recognition technologies is consistent with the public interest.

Academic Scientists

In 2019, a significant ethical dilemma arose in the scientific community when researchers called for the retraction of a paper published by Wiley. The paper revealed algorithms used to monitor Uighur Muslims in China through facial recognition technology. Written by Chinese researchers affiliated with the Academy of Military Medical Sciences and the Forensic Centre of the Ministry of Public Security, the publication sparked considerable controversy for its potential contribution to human rights abuses. This incident highlighted a wider movement within the scientific community to address unethical practices in facial recognition research. It prompted reflection on the ethical underpinnings of academic work in AI and computer vision, particularly in relation to the collection of large datasets of facial images without explicit consent.

The controversial nature of such research has been further highlighted by cases such as Stanford University's release of images from a webcam in a San Francisco cafe, and Duke University's distribution of video footage of students without their consent (Van Noorden, 2020b). There have also been alarming claims about the capabilities of facial recognition software, such as Harrisburg University's claim that its technology could predict criminal behaviour with significant accuracy, raising concerns reminiscent of discredited nineteenth-century physiognomy theories. The ensuing backlash against such claims, exemplified by a letter signed by thousands of academics urging publishers not to disseminate such research (Coalition for Critical Technology, 2020).

In response to growing ethical concerns, several initiatives have emerged to promote responsible conduct of facial recognition research. These include calls for researchers to avoid collaborating with organisations that engage in unethical practices, as well as a reassessment of data collection and distribution methods, and enhanced ethical reviews implemented by certain journals and academic conferences (Peng, 2020). In June 2020, the world's largest scientific computing society, the Association for Computing Machinery in New York City, called for the suspension of private and government use of facial recognition technology because of "clear biases based on ethnicity, race, gender, and other human characteristics" (ACM, 2020).

While some progress has been made in raising awareness of ethical issues in academic facial recognition research, a survey conducted by Nature revealed divergent perspectives among researchers on the ethical dimensions of their work (Van Noorden, 2020a). Although respondents indicated a growing consensus on the need for ethical scrutiny, many highlighted that challenges remain in reconciling the demand for large datasets, essential for training accurate facial recognition algorithms, with ethical imperatives such as obtaining informed consent. In addition, there is ongoing debate about the ethical implications of conducting research on vulnerable populations, particularly when informed consent may be difficult to obtain effectively. Ultimately, the discourse surrounding facial recognition research in academic circles underscores the need to balance scientific progress with ethical responsibility. While the condemnation of unethical applications of facial recognition is unanimous, the extent to which research practices themselves should be subject to ethical scrutiny and regulation remains controversial.

Big Tech Companies

How do powerful social actors, particularly big tech companies involved in the development and commercialisation of facial recognition technologies, navigate and address the controversies surrounding their use? As noted above, in May 2020, amidst the intensity of the Black Lives Matter movement, the debate surrounding facial recognition technologies peaked and received significant media attention. Public outcry over perceived bias against the black community in law enforcement's use of this technology led to widespread concern. In response to mounting pressure, many major tech companies—Google, Microsoft, IBM, and Amazon—halted facial recognition projects altogether or pledged not to sell the technology to law enforcement agencies (Heilweil, 2020).

Wen and Holweg (2023) analysed how companies have responded to public controversies over their use of facial recognition technologies and identified four main types of responses to mitigate reputational damage: improvement (improving the algorithm and dataset and restricting access to the technology), deflection (attributing responsibility to users, the police or researchers), validation (engaging stakeholders outside the company and seeking validation from the government and auditors), and pre-emption (clarifying prohibited uses and withdrawing from facial recognition). The authors found that companies' responses vary according to three strategic factors. First, the financial importance of the technology to the company, i.e., the revenue it has generated or is expected to generate. Second, the strategic importance of the technology, whether it is peripheral or core to the company's product and service offerings. Third, whether the controversial technology violates the company's stated public values, which serve as a cultural cornerstone that guides the company's actions, including its approach to controversial technologies.

The strategies used by big tech companies to manage controversies around facial recognition technologies, particularly through deflection, refinement, and validation, warrant further examination. These strategies are consistent with what Rommetveit and van Dijk (2022) described as forms of shifting digital governance which are part of a broader "techno-regulatory imaginary" through which (fundamental) rights protections become increasingly future-oriented and anticipatory and increasingly enacted, performed, and framed from within technological, organizational and standardization sites. Instead of traditional sites of regulation, such as law and ethics, enactments of rights become more material (technological standards and artefacts, risk management templates, etc.) and at the same time more deeply inscribed into the imagined-possibles of digital innovation deployed by engineers and tech companies.

Improving and Optimizing Algorithms

Big tech companies, along with developers like academic scientists and startups, employ regulatory measures through technical enhancements, primarily focusing on refining and optimizing algorithms. This enhancement process often includes integrating fundamental rights into technological frameworks, as emphasized by Rommetveit and van Dijk (2022). This trend reflects a growing inclination to address privacy and data protection concerns through technological means and standardization efforts. Strategies of improvement of technical robustness and safety, transparency, and accountability in regards to AI technologies— the same reasoning applies to the particular case of facial recognition technologies—"are those for which technical fixes can be or have already been developed" and "implemented in terms of technical solutions" (Hagendorff, 2020, p. 103).

Recognising issues of technical robustness and safety is an explicit admission of expert ignorance, error, or lack of control in AI technologies, of which facial recognition is no exception, and opens up space for the politics of "algorithm optimization" (Aradau & Blanke, 2022, p. 17) while reinforcing "strategic ignorance" (Aradau & Blanke, 2022, p. 89). In the words of sociologist Linsey McGoey, strategic ignorance refers to "any action that mobilises, manufactures, or exploits unknowns in a wider environment in order to avoid responsibility for previous actions" (McGoey, 2019, p. 3).

According to Aradau and Blanke (2022), when it comes to the need to correct errors, "explain algorithms", or openly address bias and discrimination, big tech companies portray AI technologies in general, and facial recognition in particular, as fallible but eternally evolving towards optimisation. Aradau and Blanke propose to unpack the emergence of "accountability by error" in global scenes of controversy around facial recognition technologies and their respective "politics of optimization and trust". In their view, accountability by error has been enacted when various publics have encountered algorithmic malfunctions, misrecognitions, failures, or other fallibilities: "These forms of accountability through error enact algorithmic systems as fallible but ultimately correctable, and therefore always desirable. Errors become temporary malfunctions, while the future of algorithms is that of indefinite optimization" (Aradau & Blanke, 2002, p. 171). In this context, having "better" facial recognition technologies (according to some positionings this is framed as "better algorithms") could mean collecting more and more data in order to optimise algorithms, thus obscuring logics of power and normalising extensive surveillance.

Distribution of Responsibility

In addition to improving facial recognition technologies as a strategy to address controversies and uncertainties and to protect the reputation of tech companies, another prominent strategy adopted by big tech companies, as shown by Wen and Holweg (2023), is to distribute responsibility for privacy and data governance in the area of facial recognition technologies by focusing on facilitating individuals' control over their data and privacy preferences. Companies are also adopting validation strategies to address controversies surrounding facial recognition technologies. This involves engaging with external stakeholders such as users, government bodies, and regulators to advocate for regulatory frameworks that support innovation.

This approach is consistent with Machado et al.'s (2023) broader discussions of AI ethics, which can be applied specifically to facial recognition technologies. They emphasise that the diffusion of responsibility and validation through stakeholder contributions helps to establish privacy and data governance as "normatively transversal" concepts (see also Steinhoff, 2023; van Dijk et al., 2018, p. 20). By distributing responsibility and seeking collective validation, companies can stabilise or resolve controversies without changing the fundamental operations of AI development. This integrated strategy of responsibility distribution and stakeholder engagement ultimately supports the ethical deployment of facial recognition technologies.

Regulators and Non-Governmental Organisations

The previous section on the strategies adopted by big tech companies to manage controversies over and rejection of facial recognition technologies illustrated how ethics and expertise for privacy and data protection are shifting away from traditional regulatory and legal professionals towards privacy engineers and risk assessors in information security and software development (Rommetveit & van Dijk, 2022). This leads to a fundamental question: How is the regulation of facial recognition technologies implemented and evolving? In this context, what is the effective impact of the actions of non-governmental organisations that are compromised by advocating for legal and regulatory safeguards in facial recognition technologies? How can alternative positions, attentive to claims about additional or alternative ethical and human rights issues, be heard and incorporated into the incremental politics of technological change?

In the context of the European Union, NGOs critical of the development of surveillance technologies and the expansion of police databases have raised ethical and social concerns about facial recognition, which is seen as highly intrusive and a violation of human rights. Euroactive, Statewatch, EDRi, and other NGOs have called on the European Commission, the European Parliament, and Member States to ensure that facial recognition technologies are comprehensively banned in law and practice across the European Union. However, while calls for a ban on facial recognition are increasingly heard in Europe, these technologies have been rapidly deployed at the borders of the European Union and continue to be seen as an essential tool of policing. At the same time, while the European Union has promoted what is often referred to as a "third way" for trustworthy AI in Europe—avoiding state surveillance in China and corporate surveillance in the United States (Aradau & Blanke, 2022, p. 139)—the restrictions on the use of real-time and retrospective facial recognition in the AI Act are minimal and do not apply to private companies or administrative authorities. As an earlier restriction—that such technology can only be used to combat serious cross-border crimes—has been removed from the final text of the AI act, a vague reference to the "threat" of a crime can now be sufficient to justify the use of retrospective facial recognition in public spaces. The AI Act leaves the option for member states to adopt stricter rules at the national level (Viet-Ditlmann & Aszódi, 2024), which is similar to the strategies of responsibility-sharing that we explored above when analysing big tech companies' strategies for managing controversies around facial recognition technologies and legitimising their continued development.

General Citizens

In recent years, we have seen an exponential increase in studies focusing on public attitudes towards facial recognition technologies, with the overarching message that there is a need to listen to the public's views in order to create the right conditions for a more informed public debate on the use of facial recognition technology, to understand what restrictions and safeguards are needed to meet public expectations, to review and clarify the legal framework for facial recognition, and to ensure that it keeps pace with public expectations. Typical social actors involved in consulting the public on facial recognition technologies include academic institutions that conduct studies and public consultations to gauge public attitudes and inform policy recommendations (Bragias et al., 2021; Kostka, 2023; Kostka et al., 2021, 2023; Lai & Rau, 2021; Miethe et al., 2023; Peng, 2023; Ramiro & Cruz, 2023; Ritchie et al., 2021); research institutes commissioned by governments to produce studies with the overall aim of overseeing the implementation and ethical use of facial recognition technologies (Ada Lovelace Institute, 2019; Davis et al., 2022); and research companies with commercial interests (IPSOS, 2019).

These studies generally show that public attitudes towards facial recognition technologies are nuanced. While there are concerns about the normalisation of surveillance, there's a willingness to accept these technologies if they offer tangible public benefits. In particular, people are generally open to their use by law enforcement for criminal investigations (Bragias et al., 2021), as well as in practical applications such as smartphone security and airport systems, provided that adequate safeguards are in place (Miethe et al., 2023). However, discomfort arises when the trade-off between public benefit and privacy invasion is not clear. There are calls for transparency, regulation, and ethical use, particularly in relation to police use. Concerns include privacy violations, normalisation of surveillance, lack of consent options, and ethical trustworthiness (Ada Lovelace Institute, 2019; Lai & Rau, 2021; Miethe et al., 2023).

Available studies of public attitudes to facial recognition technologies suggest that there is scepticism about commercial uses of facial recognition, particularly in the retail and HR/employment contexts, driven by mistrust in the ethical use of the technology by companies, although this can vary widely: for example, a cross-national comparison found that participants in the US were more accepting of tracking citizens, more accepting of the use of facial recognition technology by private companies, and less trusting of the police using facial recognition technology than people in the UK and Australia (Ritchie et al., 2021). Other factors may influence public attitudes towards facial recognition technologies: The more citizens lack trust in their government, the more likely they are to belong to the group of digital doubters (Kostka, 2023); while Peng showed that the acceptance of facial recognition applications is positively associated with right-wing authoritarianism and negatively with libertarianism, that is, individual liberty as a core principle guiding the organisation of social life. In addition, the acceptance of facial recognition technology varies across societies. For example, respondents in China generally exhibit a high level of acceptance of facial recognition technology and other surveillance measures. However, it is not entirely clear how much this acceptance is influenced by the respondents' experience of living in an authoritative state (Kostka, 2023; Kostka et al., 2021, 2023).

Despite surveys of public attitudes towards facial recognition technologies highlighting concerns about accuracy, bias, and database security, a majority of the public see facial recognition as a potential asset for public safety (Davis et al., 2022). There is a demand for transparency about data use and storage to alleviate privacy concerns. Public opinion thus reflects a balancing act between security aspirations and privacy concerns, emphasising the need for responsible development, regulation and transparent engagement with the wider community (Ramiro & Cruz, 2023). From the perspective of critical scholarship, these public engagement measures should be seen as strategies for mitigating or suppressing public controversy. This approach views the public as a valuable source of knowledge, essential for contributing to innovation, driven by a desire for legitimacy and securing public support for scientific and technological advances (Weingart et al., 2021). Engaging with public opinion on the ethical challenges of AI can also serve as a risk prevention measure, reducing conflict and closing important debates in contentious areas (Wilson, 2022).

The underlying motives for public participation raise the question of the types of publics involved, i.e., who are the publics that should be consulted in relation to facial recognition technologies. According to Machado et al.'s (2023) study of the publics involved in the debate on the ethics of AI, the general public predominates, followed by professional groups and developers of AI systems. The involvement of the general public in technological change suggests an avoidance of issues of democratic representation (Hagendorff, 2020; Schiff, 2024). Moreover, the non-specificity of "the public" does not prescribe any particular action that can secure the legitimacy and protect the interests of a wide range of stakeholders, while running the risk of silencing the voices of the very publics whose participation is sought.

The focus on the public's views on the ethical challenges of facial recognition through the general public also shows how the attempt to "lay" people's opinions can be driven by a desire to promote public trust and acceptance of facial recognition developments, and how big tech companies and other powerful actors, such as regulators and policymakers, negotiate challenges and reassert their authority. This corresponds to what Schiff (2024) has described as a new governance paradigm that encourages public participation in defining the problems and solutions associated with AI technologies that policymakers should address, but through which these calls for public engagement in AI policy are under-realised and limited by strategic considerations, where public influence only occurs when the public discusses the economic dimensions of AI.

Conclusion

This chapter showed how ethical assemblages emerging and being enacted social practices, investments, resistance to facial recognition technologies take shape through values, discourses, interactions, and practices within specific networks, deeply intertwined with socio-material and symbolic orders. In line with Michel Foucault's (1980) perspective, surveillance is power but power is never simply possessed by one individual or group and wielded over another. Power is better understood as a set of forces that have productive capacities, that are in constant flux, and that manifest in larger assemblages of material, social, and symbolic relationships which are glue holding connections in place long enough to yield social truths. One form of these assemblages being enacted is through ethical issues, being constructed and circulating through and within dynamic and interconnected networks. In this concluding section, we suggest two main strategies on how to approach ethical assemblages and networks emerging and enacted with, through and within controversies and uncertainties surrounding facial recognition technologies. These suggestions are meant to contribute to Boenig-Liptsin's (2022) proposal that we need to be more aware of the place of ethics in relation to self, community, and technology in order to make ethics a more powerful tool for understanding, questioning and shaping data and computing in the world (p. 4).

A first suggestion regards the need to move beyond binary divisions between positive and negative aspects of facial recognition technologies and their empowering and disempowering effects, and instead recognize the context-dependent place from where claims are produced (Haraway, 1988). In this sense, more research is needed to understand how facial recognition technologies are intertwined with visions of the good and evil but also, perhaps even more importantly, between and beyond binary divisions. In other words, more empirical and theoretical elaboration is needed to look across the diverse ways of being implicated and affected by facial recognition technologies, rather than considering ethical assemblages in binary terms of good and evil, inclusion versus discrimination, or technical optimism versus technological dystopian fears.

A second aspect relates to the need of broadening the analysis of facial recognition technologies in terms of the individuals, social groups, or institutions involved. Rather than seeing attempts to deal with controversies and uncertainties over facial recognition as confined to networks of innovation involving big tech companies and some academia, along with their co-opted stakeholders, we argue that these networks are evolving towards greater hybridity and boundary fusion (Machado et al., 2023). These networks, as forms of shifting digital governance, are enacted, performed, and framed in many multiple ways, while, at the same time, there is the need for a more nuanced understanding of how they operate in different social and political contexts. In this chapter, we aimed to capture different positionings and to advance the understanding of the interactions between a variety of social actors—academic scientists, big tech companies, regulators and non-governmental organizations, and citizens—each with their own status and values, and continuously reconfiguring relationships among them.

References

Abhishek, A. (2024, April 18). Thailand advances national AI strategy. Single digital ID platform to access govt services. *Biometrics News*. Retrieved May 17, 2024, from https://www.biometricupdate.com/202404/thailand-advances-national-ai-strategy

ACM. (2020). Statement on principles and prerequisites for the development, evaluation and use of unbiased facial recognition technologies. *U.S. Technology Policy Committee*. Retrieved May 21, 2024, from https://www.acm.org/binaries/content/assets/public-policy/ustpc-facial-recognition-tech-statement.pdf

Ada Lovelace Institute (2019). *Beyond face value: public attitudes to facial recognition technology*. Retrieved April 19, 2024, from https://www.adalovelaceinstitute.org/case-study/beyond-face-value/

Adjabi, I., Ouahabi, A., Benzaoui, A., & Taleb-Ahmed, A. (2020). Past, present, and future of face recognition: A review. *Electronics, 9*(8), 1188. https://doi.org/10.3390/electronics9081188

AI Sur. (2021). *Reconocimiento facial em América Latina. Tendencias de implementación de uma tecnología perversa.* Retrieved May 22, 2024, from https://www.alsur.lat/reporte/reconocimiento-facial-en-america-latina-tendencias-en-implementacion-una-tecnologia

Amnesty International. (2022). *Israel's apartheid against palestinians: Cruel system of domination and crime against humanity.* Index Number: MDE 15/5141/2022. Retrieved 29 December 2024, from https://www.amnesty.org/en/documents/mde15/5141/2022/en/Amnesty International (n/d). *Ban the scan.* Retrieved May 4, 2024, from https://banthescan.amnesty.org/

Amoore, L. (2020). *Cloud ethics: Algorithms and the attributes of ourselves and others.* Duke University Press.

Amoore, L., & Hall, A. (2010). Border theatre: On the arts of security and resistance. *Cultural Geographies, 17*(3), 299–319. https://doi.org/10.1177/1474474010368604

Andrejevic, M., Fordyce, R., Li, L., & Trott, V. (2020). *Australian attitudes to facial recognition: A national survey.* Monash University. Retrieved May 4, 2024, from https://www.monash.edu/__data/assets/pdf_file/0011/2211599/Facial-Recognition-Whitepaper-Monash,-ASWG.pdf

Aradau, C., & Blanke, T. (2022). Algorithmic reason: The new government of self and others. *Oxford University Press.* https://doi.org/10.1093/oso/9780192859624.001.0001

Bakir, V., & McStay, A. (2022). *Optimising emotions, incubating falsehoods. How to protect the global civic body from disinformation and misinformation.* Palgrave MacMillan. https://link.springer.com/book/10.1007/978-3-031-13551-4

Bauer, R. (2023). Taiwan's vast surveillance infrastructure: An achilles' heel if China invades. *Modern War Institute.* Retrieved August 28, 2024, from https://mwi.westpoint.edu/taiwans-vast-surveillance-infrastructure-an-achilles-heel-if-china-invades/

Big Brother Watch. (2018). *Face off report.* Retrieved May 16, 2024, from https://bigbrotherwatch.org.uk/campaigns/stop-facial-recognition/report/

Bigo, D., Isin, E., & Ruppert, E. (2019). *Data politics. Worlds, subjects, rights.* Routledge.

Bischoff, P. (2022). *Facial Recognition Technology (FRT): Which countries use it? [100 analyzed].* Comparitech. Retrieved April 30, 2024, from https://www.comparitech.com/blog/vpn-privacy/facial-recognition-statistics/

Black, D. (2023). Facial analysis: Automated surveillance and the attempt to quantify emotion. *Information, Communication & Society, 26*(7), 1438–1451. https://doi.org/10.1080/1369118X.2021.2011948

Bledsoe, W. (1964). *The model method in facial recognition* (Technical report). Panoramic Research, Inc.: Palo Alto, CA, USA.

Bledsoe, W., & Chan, H. (1965). *A man-machine facial recognition system. Some preliminary results.* Technical Report PRI 19A, Panoramic Research, Inc., Palo Alto, California.

Bradford, B., Yesberg, J., Jackson, J., & Dawson, P. (2020). Live facial recognition: Trust and legitimacy as predictors of public support for police use of new technology. *British Journal of Criminology.* https://doi.org/10.1093/bjc/azaa032

Bragias, A., Hine, K., & Fleet, R. (2021). 'Only in our best interest, right?' Public perceptions of police use of facial recognition technology. *Police Practice and Research, 22*(6), 1637–1654. https://doi.org/10.1080/15614263.2021.1942873

Brooks, A., & Griffiths, M. (2024). Beyond apartheid Israel. *Political Geography, 114,* 103193. https://doi.org/10.1016/j.polgeo.2024.103193

Brown, T., Statman, A., & Sui, C. (2021). *Public debate on facial recognition technologies in China.* MIT case studies in social and ethical responsibilities of computing. Retrieved May 27, 2024, from https://doi.org/10.21428/2c646de5.37712c5c

Bueno, C. (2020). The face revisited: Using Deleuze and Guattari to explore the politics of algorithmic face recognition. *Theory, Culture & Society, 37*(1), 73–91. https://doi.org/10.1177/0263276419867752

Buolamwini, J., & Gebru, T. (2018). Gender shades: intersectional accuracy disparities in commercial gender classification. *Conference on Fairness, Accountability, and Transparency. Proceedings of Machine Learning Research, 81,* 1–15. https://proceedings.mlr.press/v81/buolamwini18a/buolamwini18a.pdf

Buolamwini, J. (2023). *Unmasking AI: My mission to protect what is human in a world of machines.* Random House.

Cabitza, F., Campagner, A., & Mattioli, M. (2022). The unbearable (technical) unreliability of automated facial emotion recognition. *Big Data & Society, 9*(2), 1–17. https://doi.org/10.1177/20539517221129549

CBP (2024). *Say hello to the new face of efficiency, security and safety.* U.S. Customs and Border Protection. Retrieved June 17, 2024, from https://www.cbp.gov/travel/biometrics

Caeiro, C. (2022). *Regulating facial recognition in Latin America. Policy lessons from police, surveillance in Buenos Aires and São Paulo.* Research Paper. US and the Americas Programme. Chatham House, the Royal Institute of International Affairs. Retrieved May 9, 2024, from https://www.chatha mhouse.org/sites/default/files/2022-11/2022-11-11-regulating-facial-recogn ition-in-latin-america-caeiro.pdf

Chen, W., & Wang, M. (2023). Regulating the use of facial recognition technology across borders: A comparative case analysis of the European Union, the United States, and China. *Telecommunications Policy, 47*(2), 102482. https://doi.org/10.1016/j.telpol.2022.102482

Coalition for Critical Technology. (2020). *Abolish the #TechToPrisonPipeline. Crime prediction technology reproduces injustices and causes real harm.* Retrieved August 24, 2024, from https://medium.com/@CoalitionForCritic alTechnology/abolish-the-techtoprisonpipeline-9b5b14366b16

Dauvergne, P. (2022). *Identified, tracked, and profiled: the politics of resisting facial recognition technology.* Edward Elgar Publishing.

Davis, N., Perry, L., & Santow, E. (2022). *Facial recognition technology: Towards a model law.* Human Technology Institute, the University of Technology Sydney. Retrieved June 3, 2024, from https://www.uts.edu.au/human-tec hnology-institute/projects/facial-recognition-technology-towards-model-law

de Leeuw, K., & Bergstra, J. (2007). *The history of information security: A comprehensive handbook.* Elsevier.

del Bosque, M. (2023, February 8). Facial recognition bias frustrates black asylum applicants to US, advocates say. *The Guardian.* https://www.thegua rdian.com/us-news/2023/feb/08/us-immigration-cbp-one-app-facial-recogn ition-bias

ECNL (2023). *Technology and counter-terrorism: mapping the impact of biometric surveillance and social media platforms on civic space.* Retrieved June 17, 2024, from https://ecnl.org/publications/ct-and-tech-mapping-impact-bio metric-surveillance-and-social-media-platforms-civic

EDPB (2023, May 17). Guidelines 05/2022 on the use of facial recognition technology in the area of law enforcement. *European data protection board.* Retrieved May 22, 2024, from https://www.edpb.europa.eu/our-work-tools/our-documents/guidelines/guidelines-052022-use-facial-recognition-technology-area_en

Ellerbrok, A. (2011). Playful biometrics: Controversial technology through the lens of play. *Sociological Quarterly, 52*(4), 528–547. https://doi.org/10.1111/j.1533-8525.2011.01218.x

Eneman, M., Ljungberg, J., Raviola, E., & Rolandsson, B. (2022). The sensitive nature of facial recognition: Tensions between the Swedish police and regulatory authorities. *Information Polity, 27*, 219–232. https://doi.org/10.3233/IP-211538

Field, S. (2023, September 28). Facial recognition is everywhere—But Australia's privacy laws are 'falling way behind'. *Forbes Australia*. Retrieved May 20, 2024, from https://www.forbes.com.au/news/innovation/facial-rec ognition-is-everywhere-but-australias-privacy-laws-are-falling-way-behind/

Flores, N. (2021, November 11). En México operan tres sistemas de vigi-lancia con reconocimiento facial. *Contralínea*. Retrieved June 6, 2024, from https://contralinea.com.mx/interno/semana/en-mexico-operan-tres-sis temas-de-vigilancia-con-reconocimiento-facial/

Foucault, M. (1980). *Power/knowledge: Selected interviews and other writings, 1972–1977*. Harvester Press.

Fouquet H. (2019, October 3). France set to roll out nationwide facial recognition ID program. *Bloomberg*. Retrieved May 23, 2024, from https://www.bloomberg.com/news/articles/2019-10-03/french-liberte-tested-by-nationwide-facial-recognition-id-plan

Fry, H. (2018). *Hello world: How to be human in the age of the machine*. W. W. Norton & Company.

Galligan, C., Rosenfeld, H., Kleinman, M., & Parthasarathy, S. (2020). *Cameras in the classroom: Facial recognition technology in schools (tech-nology assessment report)*. Gerald R. Ford School of Public Policy. Science, technology, and public policy, University of Michigan. Retrieved April 16, 2024, from https://stpp.fordschool.umich.edu/sites/stpp/files/uploads/file-assets/cameras_in_the_classroom_full_report.pdf

Gates, K. (2011). *Our biometric future: Facial recognition technology and the culture of surveillance*. New York University Press.

Gentile, G. (2023). *Does big brother exist? Face recognition technology in the United Kingdom* (LSE Law, Society and Economy Working Papers, 7/2023). https://papers.ssrn.com/sol3/papers.cfm?abstract_id=4394694

Greenfield, A. (2017). *Radical technologies: The design of everyday life*. Verso.

Gupta, G., Raja, K., Gupta, M., Jan, T., et al. (2023). A comprehen-sive review of DeepFake detection using advanced machine learning and fusion methods. *Electronics, 13*(1), 95. https://doi.org/10.3390/electronics1 3010095

Hagendorff, T. (2020). The ethics of AI ethics: An evaluation of guidelines. *Minds and Machines, 30*, 99–120. https://doi.org/10.1007/s11023-020-095 17-8

Haraway, D. (1988). Situated knowledges: The science question in feminism and the privilege of partial perspective. *Feminist Studies, 14*(3), 575–599.

Hao, K., & Swart, H. (2022, April 19). South Africa's private surveillance machine is fueling a digital apartheid. *MIT Technology Review*. Retrieved

May 23, 2024, from https://www.technologyreview.com/2022/04/19/104 9996/south-africa-ai-surveillance-digital-apartheid/

Heilweil, R. (2020, June 11). Big tech companies back away from selling facial recognition to police. That's progress. *Recode.* Retrieved April 11, 2024, from https://www.vox.com/recode/2020/6/10/21287194/ama zon-microsoft-ibm-facial-recognition-moratorium-police.

Hupont, I., Tolan, S., Gunes, H., & Gomez, E. (2022). The landscape of facial processing applications in the context of the European AI act and the development of trustworthy systems. *Scientific Reports, 12*(10688), 1–14. https://doi.org/10.1038/s41598-022-14981-6

ICO—Information Commissioners' Office (2021, June 18). *Information commissioner's opinion. The use of live facial recognition technology in public places.* Retrieved April 18, 2024, from https://ico.org.uk/media/2619985/ ico-opinion-the-use-of-lfr-in-public-places-20210618.pdf

Instituto Garapé. (2023). *Reconhecimento facial no Brasil.* Retrieved April 17, 2024, from https://igarape.org.br/infografico-reconhecimento-facial-no-bra sil/

Introna, L., & Wood, D. (2004). Picturing algorithmic surveillance: The politics of facial recognition systems. *Surveillance & Society, 2*(2/3), 177–198. https://doi.org/10.24908/ss.v2i2/3.3373

IPSOS (2019). *Global public opinion on government use of AI and facial recognition.* An Ipsos Survey for the World Economic Forum. Retrieved April 22, 2024, from https://www.ipsos.com/sites/default/files/ct/news/doc uments/2019-09/wef-global-public-opinion-on-government-use-of-facial-recognition.pdf

Jackson, A. (2024, January 15). A facial-recognition tour of New York. *The New Yorker.* Retrieved April 23, 2024, from https://www.newyorker.com/ magazine/2024/01/22/a-facial-recognition-tour-of-new-york

Jasanoff, S., Markle, G., Petersen, J., & Pinch, T. (Eds.). (1995). *Handbook of science and technology studies.* Sage Publications.

Kassenova N., & Duprey, B. (Eds.). (2021). *Digital silk road in Central Asia: Present and future.* Davis Center for Russian and Eurasian Studies: Harvard University. Retrieved June 11, 2024, from https://daviscenter.fas.harvard. edu/research-initiatives/program-central-asia/digital-silk-road-central-asia-present-and-future

Kaur, P., Krishan, K., Sharma, S., & Kanchan, T. (2020). Facial-recognition algorithms: A literature review. *Medicine, Science and the Law, 60*(2), 131–139. https://doi.org/10.1177/0025802419893168

Kloppenburg, S., & van der Ploeg, I. (2018). Securing identities: Biometric technologies and the enactment of human bodily differences. *Science as Culture, 29*(1), 57–76. https://doi.org/10.1080/09505431.2018.1519534

Kostka, G. (2023). Digital doubters in different political and cultural contexts: Comparing citizen attitudes across three major digital technologies. *Data & Policy., 5*, e27. https://doi.org/10.1017/dap.2023.25

Kostka, G., Steinacker, L., & Meckel, M. (2021). Between security and convenience: Facial recognition technology in the eyes of citizens in China, Germany, the United Kingdom, and the United States. *Public Understanding of Science, 30*(6), 671–690. https://doi.org/10.1177/09636625211001555

Kostka, G., Steinacker, L., & Meckel, M. (2023). Under big brother's watchful eye: Cross-country attitudes toward facial recognition technology. *Government Information Quarterly, 40*(1). https://doi.org/10.1016/j.giq.2022.101761

Lai, X., & Rau, P. (2021). Has facial recognition technology been misused? A public perception model of facial recognition scenarios. *Computers in Human Behavior, 124*, 106894. https://doi.org/10.1016/j.chb.2021.106894

Leslie, D. (2020). *Understanding bias in facial recognition technologies: An explainer*. The Alan Turing Institute. Retrieved May 14, 2024, from https://doi.org/10.5281/zenodo.4050457

Lyon, D. (2010). Identification, surveillance and democracy. In K. D. Haggerty & M. Samatas (Eds.), *Surveillance and democracy* (pp. 34–50). Routledge.

Macdonald, A. (2023, October 16). Taiwanese pols raise security concerns with biometric gear from China, Thailand. *Biometric News*. Retrieved May 15, 2024, from https://www.biometricupdate.com/202310/taiwanese-pols-raise-security-concerns-with-biometric-gear-from-china-thailand

Machado, H., Silva, S., & Neiva, L. (2023). Publics' views on ethical challenges of artificial intelligence: A scoping review. *AI Ethics*. https://doi.org/10.1007/s43681-023-00387-1

Magnet, S. (2011). *When biometrics fail: Gender, race, and the technology identity*.

Martínez, J. (2022). The dark side of progress: Social and political movements against artificial intelligence in Spain. In J. Saura & F. Debasa (Eds.), *Handbook of research on artificial intelligence in government practices and processes* (pp. 226–242). IGI Global.

McConvey, J. (2024a, February 12). Expansive facial recognition surveillance coming to Hong Kong, Bahrain, South Africa. *Biometric News*. Retrieved

May 15, 2024, from https://www.biometricupdate.com/202402/expansive-facial-recognition-surveillance-coming-to-hong-kong-bahrain-south-africa

McConvey, J. (2024b, May 8). Thai government to collect iris and face biometrics from Myanmar nationals. *Biometric News*. Retrieved May 17, 2024, from https://www.biometricupdate.com/202405/thai-government-to-collect-iris-and-face-biometrics-from-myanmar-nationals

McGoey, L. (2019). *The unknowers. How strategic ignorance rules the word*. Zed.

Metz, R. (2021, November 8). Facebook is shutting down its facial recognition software. *CNN Business*. Retrieved June 4, 2024, from https://edition.cnn.com/2021/11/02/tech/facebook-shuts-down-facial-recognition/index.html

Milmo, D. (2022, May 23). UK watchdog fines facial recognition firm £7.5m over image collection. *The Guardian*. Retrieved May 6, 2024, from https://www.theguardian.com/technology/2022/may/23/uk-data-watchdog-fines-facial-recognition-firm-clearview-ai-image-collection

Miethe, T., Dudinskaya, T., Forepaugh, C., & Sousa, W. H. (2023). Facial recognition technology in policing: A national survey of public support for this technology and privacy/safety concerns. *Crime & Delinquency*. https://doi.org/10.1177/00111287221150172

Mozur, P. (2019, April 14). One month, 500,000 face scans: How China is using A.I. to profile a minority. *The New York Times*. Retrieved June 6, 2024, from https://www.nytimes.com/2019/04/14/technology/china-surveillance-artificial-intelligence-racial-profiling.html

Nash, J. (2023, March 17). SuperCom picks up last phase of suspect-monitoring project in Romania. *Biometric News*. Retrieved June 7, 2024, from https://www.biometricupdate.com/202303/supercom-picks-up-last-phase-of-suspect-monitoring-project-in-romania

Nash, J. (2023, October 4). Gang violence pushes Sweden to consider face biometrics as surveillance cameras multiply. *Biometric News*. Retrieved May 8, 2024, from https://www.biometricupdate.com/202310/gang-violence-pushes-sweden-to-consider-face-biometrics-as-surveillance-cameras-multiply

National Academies. (2024). *Facial recognition technology: Current capabilities, future prospects, and governance* (Consensus Study Report). The National Academies Press. Retrieved May 22, 2024, from https://doi.org/10.17226/27397

Nikolic, I. (2020, February 28). French court rules against facial recognition in high schools. *B.I.R.D.*. Retrieved May 3, 2024, from https://bird.tools/french-court-rules-against-facial-recognition-in-high-schools/

NYPD - New York Police Department (2023). *Facial recognition: Impact and use policy*. Retrieved May 6, 2024, from https://www.nyc.gov/assets/nypd/downloads/pdf/public_information/post-final/facial-recognition-nypd-impact-and-use-policy_4.9.21_final.pdf

O.D.I.A. (2022). *Observatorio de derecho informatico argentino O.D.I.A. y otros contra gcba sobre amparo – otros*. Juzgado en lo Contencioso Administrativo y Tributario Nº 4 de la Ciudad Autónoma de Buenos Aires. Retrieved June 4, 2024, from https://repositorio.mpd.gov.ar/jspui/handle/123456789/4658

Pearson, J. (2024, May 2). The breach of a face recognition firm reveals a hidden danger of biometrics. *Wired*. Retrieved June 4, 2024, from https://www.wired.com/story/outabox-facial-recognition-breach/

Peng, K. (2020). Facial recognition datasets are being widely used despite being taken down due to ethical concerns. Here's how. *Freedom to tinker. Research and commentary on digital technologies in public life*. Retrieved May 8, 2024, from https://freedom-to-tinker.com/2020/10/21/facial-recognition-datasets-are-being-widely-used-despite-being-taken-down-due-to-ethical-concerns-heres-how/

Peng, Y. (2023). The role of ideological dimensions in shaping acceptance of facial recognition technology and reactions to algorithm bias. *Public Understanding of Science, 32*(2), 190–207. https://doi.org/10.1177/09636625221113131

Polisen (2024). *Facial recognition in airport*. Retrieved June 4, 2024, from https://polisen.se/en/laws-and-regulations/travel-to-and-stay-in-sweden/border-control/bearer-verification/

Pugliese, J. (2010). Bodies, technologies, biopolitics. *Routledge*. https://doi.org/10.4324/9780203849415

Ramiro, A., & Cruz, L. (2023). The grey-zones of public-private surveillance: Policy tendencies of facial recognition for public security in Brazilian cities. *Internet Policy Review. Journal of Internet Regulation, 12*(1), 1–23.

Ritchie, K., Cartledge, C., Growns, B., et al. (2021). Public attitudes towards the use of automatic facial recognition technology in criminal justice systems around the world. *PLoS ONE, 16*(10), e0258241. https://doi.org/10.1371/journal.pone.0258241

Rommetveit, K., & Van Dijk, N. (2022). Privacy engineering and the techno-regulatory imaginary. *Social Studies of Science, 52*(6), 853–877. https://doi.org/10.1177/03063127221119424

Sánchez-Monedero, J., & Dencik, L. (2022). The politics of deceptive borders: 'Biomarkers of deceit' and the case of iBorderCtrl. *Information, Communication & Society, 25*(3), 413–430. https://doi.org/10.1080/1369118X.2020.1792530

Schiff, D. (2024). Framing contestation and public influence on policymakers: Evidence from US artificial intelligence policy discourse. *Policy and Society*, puae007. https://doi.org/10.1093/polsoc/puae007

Smith, M., & Miller, S. (2022). The ethical application of biometric facial recognition technology. *AI & Society, 37*, 167–175. https://doi.org/10.1007/s00146-021-01199-9

Smith, M., & Miller, S. (2021). Facial recognition and privacy rights. In *Biometric identification, law and ethics* (pp. 21–28). Springer. https://doi.org/10.1007/978-3-030-90256-8_2

Søraa, R. (2023). *AI for diversity*. Routledge.

Steinhoff, J. (2023). AI ethics as subordinated innovation network. *AI & Society*, 1–13. https://doi.org/10.1007/s00146-023-01658-5

Strong, M. (2023, October 13). Taipei city accused of using biometric recognition systems made in China. *Taiwan News*. Retrieved June 5, 2024, from https://www.taiwannews.com.tw/news/5019609

Stryker, C. (2021). Digital silk road and surveillance technology in central Asia. In N. Kassenova & B. Duprey (Eds.), *Digital silk road in central Asia: Present and future* (pp. 17–54). Davis Center for Russian and Eurasian Studies.

Tiremeurostodasuamira. (2022, June 21). *Minha operadora. Ativistas pedem banimento do reconhecimento facial na segurança pública*. Retrieved April 19, 2024, from https://tiremeurostodasuamira.org.br/minha-operadora-ativistas-pedem-banimento-do-reconhecimento-facial-na-seguranca-publica/

Tyler, H. (2022, February 2). The increasing use of artificial intelligence in border zones prompts privacy questions. *Migration Policy Institute*. Retrieved April 12, 2024, from https://www.migrationpolicy.org/article/artificial-intelligence-border-zones-privacy

van Dijk, N., Tanas, A., Rommetveit, K., & Raab, C. (2018). Right engineering? The redesign of privacy and personal data protection. *International Review of Law, Computers and Technology, 32*(2–3), 230–256. https://doi.org/10.1080/13600869.2018.1457002

Van Noorden. (2020a, November 18). The ethical questions that haunt facial-recognition research. *Nature—News Feature*. Retrieved May 9, 2024, from https://www.nature.com/articles/d41586-020-03187-3

Van Noorden. (2020b, November 19). What scientists really think about the ethics of facial recognition research. *Nature—News Feature*. Retrieved May 9, 2024, from https://www.nature.com/articles/d41586-020-03257-6

Vavoula, N. (2020). *Police information exchange. The future developments regarding Prüm and the API Directive*. Study commissioned for the LIBE Committee. Retrieved April 12, 2024, from https://www.europarl.europa.eu/thinktank/en/document/IPOL_STU(2020)658542

Viet-Ditlmann, K., & Aszódi, N. (2024, March 13). EU parliament votes on AI Act; Member states will have to plug surveillance loopholes. *Algorithm Watch*. Retrieved May 3, 2024, from https://algorithmwatch.org/en/eu-parliament-votes-on-ai-act/

Weingart, P., Joubert, M., & Connoway, K. (2021). Public engagement with science—Origins, motives and impact in academic literature and science policy. *PLoS ONE, 16*(7), e0254201. https://doi.org/10.1371/journal.pone.0254201

Wen, Y., & Holweg, M. (2023). A phenomenological perspective on AI ethical failures: The case of facial recognition technology. *AI & Society*. https://doi.org/10.1007/s00146-023-01648-7

West, S. M. (2019). Data capitalism: Redefining the logics of surveillance and privacy. *Business & Society, 58*(1), 20–41. https://doi.org/10.1177/0007650317718185

Wilson, C. (2022). Public engagement and AI: A values analysis of national strategies. *Government Information Quarterly, 39*(1), 101652. https://doi.org/10.1016/j.giq.2021.101652

3

The Evolving Scenarios of Artificial Intelligence in Assisted Reproductive Technologies

Abstract This chapter explores the often-futuristic potential of artificial intelligence in the context of assisted reproductive technologies, and the unique array of ethical and practical dilemmas that emerge in the academic literature, online discourses, and bio-reproductive platforms, such as embryo and patient data repositories and research collaboration networks, that are used by providers of assisted reproductive services. It begins by critically examining a range of positions on the current controversies and uncertainties surrounding assisted reproductive technologies, including issues such as ethical boundaries, access and inequality, consent and informed decision-making, regulatory oversight, expected outcomes, effectiveness and clinical validation, and technological evolution. The chapter also discusses future pathways and ongoing efforts to promote ethics in the field of embryo selection. Through the lens of assemblage theory, it explores how complex networks are negotiating ethical principles and setting standards to shape evolving collective understandings of what is considered good practices that generate excitement and mobilise resources for clinical, biotechnological, and commercial applications. We challenge the dominance of experts and technocratic approaches, arguing for a nuanced understanding of ethical dilemmas as relational and contextual.

© The Author(s), under exclusive license to Springer Nature
Singapore Pte Ltd. 2025
H. Machado and S. Silva, *Ethical Assemblages of Artificial Intelligence*,
https://doi.org/10.1007/978-981-96-4158-1_3

Keywords Assisted reproductive technologies · Ethical assemblages · Artificial intelligence · Ethical dilemmas · Controversies · Uncertainties · Embryo selection

Introduction

The application of artificial intelligence (AI) in assisted reproductive technologies (ART) entails the utilisation of machine-learning algorithms, computer vision, and robotics to manage, process, and analyse large, temporal and dynamic datasets generated throughout ART cycles. The primary goal is enhancing the success rates of ART. In particular, AI promises to improve the effectiveness of different stages of fertility treatments by analysing datasets comprising images, genetic, morphological, and clinical information, thereby contributing to the optimisation and personalisation of care (Hanassab et al., 2024). This chapter examines the often-futuristic potential of AI in the context of ART, illustrating how AI could lead to profoundly transformative developments in the field of medical care. However, we also highlight the unique array of controversies, uncertainties, ethical concerns, and practical challenges that need to be addressed.

Let us consider a scenario that comes closest to the promises of AI's ability to assist with fertility treatments, as envisioned by developers and experts advocating for these technologies. Firstly, AI is seen as capable of relentlessly sifting through vast amounts of data from countless previous ART cycles to identify patterns in the critical factors that influence success rates, with supposedly unprecedented accuracy. Through meticulous data analysis, supported by machine-learning algorithms, AI is expected to provide fertility doctors with deeper insights and build robust predictive models needed to support decision-making and tailor treatments to each patient's unique and specific circumstances. Second, AI is seen as having the potential to facilitate the complicated process of embryo selection. Traditional methods depend significantly on human visual assessment, whereas machine-learning algorithms can analyse detailed images and complex datasets to predict which embryos are most likely to result in a successful pregnancy.

In addition, AI can enhance sperm analysis by assessing sperm motility and morphology, ensuring that only the most suitable candidates are selected for fertilisation. This innovative approach is expected to demonstrate superior performance compared to traditional methods, with the potential to increase the likelihood of success. AI is also expected to outperform traditional methods in monitoring and prediction. It is expected to improve continuous monitoring of patients' responses to hormone treatments and to predict optimal times for critical procedures such as egg retrieval and embryo transfer. Finally, the ability of AI to personalise treatment plans is being hailed by fertility experts and clinics as a living proof of its transformative power in ART. The integration and analysis of data specific to each patient is expected to enable the creation of bespoke treatment strategies through the use of AI. This personalised approach is seen by developers of AI in the field of ART as not only increasing the chances of success, but also reducing the time and cost involved, thereby making parenthood more accessible to many.

The application of AI in ART can be traced back to the late twentieth century (Medenica et al., 2022). The initial applications of AI in ART focused on data management and analysis, using machine-learning algorithms employed to analyse large sets of clinical data to predict outcomes and optimise procedures. However, an analysis of the academic literature revealed that it is only in the last five years that several research teams have invested in developing important applications of AI in ART, with the aim of optimising gamete efficiency and setting standards for the procedures, particularly in the field of embryo selection. Consequently, AI is still in the early stages of clinical implementation, with its applications currently being disseminated through reproductive medicine and fertility professional societies, scientific journals, and clinic websites before becoming a widespread reality. In this context, addressing the controversies, uncertainties, and ethical and practical dilemmas of AI in ART through the concept of ethical assemblages might contribute to move the debate towards an "ethics of possibility" (Appadurai, 2013), that expands collective imagination and visions of the kind of society that AI in ART can bring into being:

> Imaginaries support the shift from expectations based on past experiences (which entails the formulation of new expectations) to a more open future – that is, the shift from 'the ethics of probability' (concerned with what will probably happen) to an 'ethics of possibility' (increase the horizons of hope and expand the field of the imagination') (Appadurai, 2013: 295). (Bazzani, 2023, p. 388)

The exponential growth of clinical applications of AI in the field of ART in recent years has been described as a gold rush (Lee et al., 2024, p. 286). This is due to the sudden and rapid influx of seekers pursuing AI as the "gold" that has been discovered to improve overall ART success rates and to personalise fertility treatments. This chapter examines the controversies, uncertainties and ethical dilemmas associated with this process. We explore how complex networks are influencing the development of ethical standards and principles to shape the future of clinical, biotechnological, and commercial applications. Our focus is on the dynamic and intricate web of social relations and epistemic practices that collectively contribute to the creation, sharing, and legitimisation of knowledge within the academic, industrial, and professional realms of AI in the context of ART.

The methodological framework proposed by Aradau and Blanke (2022, p. 12), referred to as the "scene", provides a valuable lens through which to examine the controversies, uncertainties, debates, and disputes surrounding AI in ART. Our analysis focuses on the "scenes" that circulate in the academic literature, particularly in the form of editorials, commentaries, and review articles. We also examine the narratives and bio-reproductive platforms, such as embryo and patient data repositories and research collaboration networks, that are used by providers of assisted reproductive services. These platforms include AI-enhanced databases that store and analyse information, to help to refine and personalise future ART cycles. The research collaboration networks that emerge from these platforms enable data sharing and collaboration between academic researchers and fertility clinics to continuously improve AI algorithms and key steps in an ART cycle. These "scenes" are enacted and performed through specific technical, social, and political practices that exclude what is deemed irrelevant, select the key issues, interpret their value and meaning, establish a temporal order that enables causal modelling, and support action (Bazzani, 2023). Consequently, they exemplify "situated knowledges and partial perspectives" (Haraway, 1988).

This chapter proceeds as follows. The first section examines the main controversies surrounding AI in ART, including issues such as ethical boundaries, access and inequality, consent and informed decision-making, and the expected outcomes of AI applications. In the second section, we highlight the major uncertainties that arise in the specific field of ART, including regulatory oversight, ethical standards, effectiveness and clinical validation, and technological evolution, contrasting the perspectives of developers with critics.[1] In the third section, we describe ongoing efforts to promote key ethical principles and set standards within a major controversial and promising application of AI in ART—embryo selection. The fourth section explores the future pathways for AI in ART. The fifth section systematises the diverse positions of social actors from academia, industry, and clinical practice in reproductive medicine and fertility, and how they intertwine through networks to negotiate ethics and to shape unstable and evolving collective understandings of what is considered "good practices" that generate excitement and mobilise resources in the field of AI in ART. The final section summarises the role of ethical assemblages and the related material, social, and symbolic landscapes in the field of AI in ART.

---

[1] Those involved in the development and promotion of AI in ART include individuals and organisations responsible for the creation, design, and implementation of AI technologies for use in ART. Their aim is to innovate and enhance AI applications to improve the efficiency of clinical procedures and the success rates of ART. Critics of AI in ART include individuals or groups who analyse, evaluate, and question the use and impact of AI technologies in this field. These actors place particular emphasis on the ethical, social, and clinical implications of the technologies in question, and advocate for rigorous validation and responsible implementation.

Controversies of Artificial Intelligence in Assisted Reproductive Technologies

The introduction of AI into the field of ART has not been without controversies. Table 3.1 outlines the main concerns related to AI in ART, focusing on ethical boundaries in human reproduction, access and inequality, consent and informed decision-making, and expected outcomes. The subsequent sections will provide a detailed examination of each of these controversies.

Ethical Boundaries

Setting ethical boundaries for AI intervention in human reproduction is not happening in a vacuum. It is influenced by cultural, social, and political factors that shape views of what is acceptable and desirable. A

Table 3.1 Controversies surrounding AI in assisted reproductive technologies

Controversy category	Main elements
Ethical boundaries	• Frontiers of intervention in human reproduction • Implicit morality of the algorithms • Potential responsibility gaps
Access and inequality	• Concerns about equitable benefit • Potential for exacerbating social inequalities in access to reproductive healthcare • Emergence of new social pressures for using AI in ART
Consent and informed decision-making	• To fully understand, explain, and consent to the use of AI applications is challenging • Concerns about "machine paternalistic" decision-making • Potential misrepresentation of patient's values
Expected outcomes	• Debates around meaningful outcomes • Concerns about what should be measured/captured and how

key question concerns the underlying moral values and ethical implications of algorithms (Amoore, 2023; Tsamados et al., 2022), particularly in relation to fairness, accountability, and "new thresholds of normality and abnormality" (Amoore, 2000, p. 6).

Consider, for example, how predictive algorithms used to select gametes and embryos are largely described as being based on objective clinical criteria and relying on data inputs that can be quantified, yet often take place within sociotechnical and cultural infrastructures, such as the datasets of the algorithms and the selection and labelling of training data, i.e., the conditions of an algorithm's emergence that perpetuate long-standing structural inequalities. This selection process includes the opacity of disability rights. The selection of gametes and embryos that are free of genetic diseases or disabilities can serve to reinforce societal prejudice and discrimination against people with these conditions. This is because such a process suggests that these people are somehow inferior or unworthy, and that these conditions are undesirable. Such circumstances could potentially influence the formation of their personal and social identities. Those born with such conditions may internalise the idea that they are devalued by society, which could have a detrimental effect on their self-esteem and sense of belonging. Socially, it can lead to stigmatisation and exclusion of these people, which could further marginalise them in society. In addition, there are concerns that AI technologies could facilitate the selection of embryos based on certain desirable non-medical characteristics, such as gender and ethnicity (Afnan et al., 2021). This could lead to a future where societal pressures to enhance humans and genetics become normalised, hindering the development of an inclusive society that fully embraces diversity and potentially perpetuating social inequalities.

AI in ART has often been criticised for the implicit moral implications of the algorithms used to select the "best" gametes or embryos, which reflect societal inequalities and deeply held values about life, health, disability, and societal norms. Additionally, there are concerns about who holds moral responsibility and accountability for the selection process, especially when the design, development, and use of these algorithms are distributed across multiple actors, making traceability difficult. This difficulty, defined as the challenge of identifying and tracing the chain of events and factors contributing to a particular selection by distributed and hybrid systems, that include human and non-human actors, can lead to a responsibility gap. This gap potentially erodes ethical and legal accountability (Afnan et al., 2021) and results in "agency laundering", where the algorithm is blamed for morally questionable actions (Rubel et al., 2019).

Access and Inequality

One of the key challenges is to ensure equitable access and benefit in a market that is characterised by a shortage of funding and access restrictions in most countries, alongside the increasing ownership of fertility clinics by private investment firms (Perrotta, 2024). This trend can influence how the clinics operate, potentially focusing on profitability and returns on investment, which can impact equitable access to fertility care and the quality of services provided. In a context where access to AI in the field of ART is constrained by socioeconomic status, social inequalities might exacerbate. Furthermore, only those with higher digital literacy may be able to effectively use and manipulate these algorithms, further exacerbating social inequality (Martin, 2019).

The vision of AI in ART as a solution bridging current low success rates with future advancements in human reproduction helps drive the expansion of these technologies. This vision may also create new social pressures to use AI in ART, especially within a new business model focused on future-oriented, speculative value creation and market expansion driven by disruptive technologies (Hogarth, 2017; Patrizio et al., 2022).

Consent and Informed Decision-Making

Both clinicians and patients find it difficult to fully understand AI models and explain their decision-making processes due to the complexity and opacity of AI in ART. This challenge is exacerbated by the information asymmetry between the companies selling these tools and the clinicians who must make daily decisions using them (Afnan et al., 2021, 2022). This information asymmetry might compromise shared decision-making and create an opportunity for a "machine paternalistic" decision-making process (Afnan et al., 2021), where AI systems make choices on behalf of clinicians and patients, presuming to know what is best for them, potentially misrepresenting their values.

The introduction of AI into decision-making processes on behalf of patients, clinicians, and embryologists, ostensibly for their own good and in their best interests, but without their explicit consent or against their preferences, raises significant ethical and practical concerns. These concerns relate to the potential loss of autonomy, lack of accountability and/or oversight, trust issues, and the potential for abuse by leaving end-users in a permanent state of confusion and destabilisation so that they do not question an algorithm's suggestion (Tsamados et al., 2022). AI in ART can sometimes override the choices of individuals and their right to make their own decisions. In the event of harm resulting from decisions made by AI, it is imperative to determine who bears responsibility for such decisions.

Furthermore, AI often requires access to personal data, which raises concerns not only about privacy and security, but also about the full informed consent of those involved in retrospective studies, it is, research using past data to develop AI tools for ART. This means that participants must be fully aware of how their data will be used and agree to it. Variations in regulations about what constitutes adequate informed consent further complicate the situation.

An illustrative example is a multi-centre cohort study using data from 8147 biopsied blastocysts from 1725 patients treated at nine UK IVF clinics between 2012 and 2020 (Bamford et al., 2023), in which it was decided that consent was not required for participation. The decision was based on the assumption that neither patient care nor the outcome for the embryos would be impacted (Bamford et al., 2023). However, it failed to consider the autonomy and preferences of the patients involved.

This study did not require ethical approval as it utilized anonymized, numerical data acquired during routine, validated and HFEA-licenced practices. The data analyses were approved by the participating clinics research and innovation board, and it was decided that consent was not necessary for participation in this retrospective data analysis as patient care and fate of the embryos was unaffected. (Bamford et al., 2023, p. 570)

The above study was exempted from ethical scrutiny on the basis of the argument that social problems in the use of AI can be avoided by keeping data separate from people (Helm et al., 2024). However, other retrospective scientific studies have taken an alternative approach. One such example is a retrospective analysis of a dataset of 11,286 cycles performed at a single centre in France between 2009 and 2020 (Ferrand et al., 2023). This study was approved by the local institutional review board committee and all participants provided written consent for their data to be used in retrospective scientific studies.

These examples highlight the inconsistencies in the regulation of the legal use of large datasets in different countries. A key issue is the varying levels of de-identification of patient information, i.e., the process of removing or altering personal information from data so that individuals cannot be easily identified. Different countries have different standards for how thoroughly data must be de-identified, which can affect the privacy and protection of patient information. Different regulations on what constitutes adequate informed consent further complicate the situation.

Expected Outcomes

The question of what constitutes a meaningful outcome generated by AI in the context of ART remains a topic of ongoing debate. In particular, there is a need to determine what the expected outcomes or endpoints of AI applications in ART should be, and how they should be measured, recorded, and documented. Outcomes are multi-faceted and encompass multiple measures, including clinical, economic, and patient-centred outcomes. For example, the birth of a live child is a clinical outcome, while resource utilisation or patient satisfaction are economic and patient-centred outcomes, respectively. Each of these outcomes provides unique and different insights into the effectiveness and impact of AI technologies.

A significant outcome highlighted by developers is the assertion that AI can outperform humans and potentially exceed human capabilities in clinical settings. They argue that AI could provide more objective, faster, and more accurate assessments compared to trained clinicians and embryologists (Salih et al., 2023). However, scholars in science and technology studies (STS) stress the need to consider the broader context and the locality of algorithms and AI (Geampana & Perrotta, 2023). First, performance metrics observed in controlled study environments may not be directly applicable to real-world clinical settings, where the social and organisational contexts in which AI operates can significantly influence its effectiveness and acceptability. Secondly, AI lacks the contextual understanding, ethical judgement, and empathy inherent in clinicians and embryologists. Consequently, monitoring the implementation, development, and use of AI applications in clinical care in the field of ART cannot be reduced to controlled performance metrics and must include contextual factors. It is imperative to investigate the previously unobserved local working practices required to implement AI-based ART in practice. Such practices often require adaptation and improvisation by users (Bowker & Star, 2000), challenging the subjective/objective dichotomy, where objectivity is associated with knowledge standardisation and certainty, while subjectivity is associated with high levels of human involvement (Geampana & Perrotta, 2023).

Developers have highlighted that improving success rates in ART is a major benefit of using AI in fertility clinics. Given the current success rates, which range from 20 to 30%, and the concerning trend of decline—partly due to the increasing average age of women seeking fertility treatment—AI in ART aims to address these challenges (Salih et al., 2023). Furthermore, with a forecasted shortage of skilled embryologists to handle the increasing demand for IVF services, AI is being positioned as a key solution to improving ART success rates (Sadeghi, 2022). However, it is important to question how success rates in ART are measured. Currently, success metrics focus on intermediate clinical outcomes such as clinical pregnancy and embryo implantation, which may not be fully aligned with patients' primary concerns (Rieger et al., 2021). For patients, the most important outcome of AI in ART is the birth of a healthy, living child. However, there is currently no peer-reviewed research evaluating how AI might significantly influence the likelihood of achieving a healthy live birth. This issue remains unaddressed despite calls from patients, clinicians, embryologists, and critical academics for AI developers to focus on predicting live births to enhance the clinical relevance of AI in ART (Salih et al., 2023). The importance of considering the birth of a healthy child as a measure of success is highlighted by academics as Chow and colleagues:

> To date, the vast majority of publications have utilised positive pregnancy test as the metric of success. It is our opinion that the most appropriate and patient-centred target is the birth of a healthy child. (…) We should evaluate AI not only by the capacity to enhance existing clinical assessments but on their ultimate improvement to real clinical targets, such as live birth. (Chow et al., 2021, pp. C22–C33)

When algorithmic systems simplify the complex nature of reproductive politics (Silva & Machado, 2011) to focus solely on primary clinical outcomes, such as a positive pregnancy test, they treat the evaluation of AI in ART as a purely technical problem. This approach reduces the broader social and political issues to mere machine-learning challenges (Amoore, 2023), thereby limiting the consideration of other important factors. For example, it overlooks the integration of patients' values into AI in ART, focusing only on optimising abstract data representations rather than addressing the full range of social and political concerns.

Uncertainties of Artificial Intelligence in Assisted Reproductive Technologies

Although developers and experts promoting AI in ART see the field as promising and offering numerous benefits, such as improving ART success rates and fertility outcomes for patients, enhancing personalised reproductive medicine, optimising gamete efficiency, increasing automation and standardisation, and reducing human subjectivity, especially in areas involving visual analysis, not everyone shares this techno-optimistic perspective. Critical scholars and some reproductive specialists, among others, have publicly expressed concerns about the uncertainties of AI in ART (e.g., Afnan et al., 2021, 2022; Chow et al., 2021; Geampana & Perrota, 2023; Riegler et al., 2021).

Table 3.2 outlines the principal uncertainties pertaining to AI in ART, with particular emphasis on the pivotal issues of regulatory oversight, ethical standards, effectiveness and clinic validation, and technological evolution. In the following sections, we provide a comprehensive examination of each of these uncertainties.

Table 3.2 Uncertainties surrounding AI in assisted reproductive technologies

Uncertainty category	Main elements
Regulatory oversight	• How AI in the field of ART should be regulated and validated • Different guidelines and recommendations • Emerging regulatory framework
Ethical standards	• Different cultural and moral perspectives • Ethical quandaries about the transparency of AI tools and data • Debates about the promotion of generalisability, fairness, and trustworthiness
Effectiveness and clinic validation	• Scarce evidence for the effectiveness of AI in improving ART success rates • Lack of clinical validation • Divide between academia and industry on clinical validation: randomised controlled trials versus larger international multi-centre non-randomised studies
Technological evolution	• Debates on implications of potential future convergence of AI in ART with in vitro gametes • Putative consequences for our idea of human reproduction

Regulatory Oversight

The lack of clarity surrounding the regulatory oversight of AI in ART has led to significant debate within professional societies for reproductive medicine and fertility, as well as ethics committees. The question of how AI in ART should be regulated and validated (Hanassab et al., 2024; Medenica et al., 2022) is a key point of contention, as is the question of whether the use of such technology is an appropriate response to improving ART success rates. The regulatory framework for AI in ART is still evolving, with varying guidelines and recommendations in different countries, creating uncertainty for clinics and patients alike.

Ethical Standards

Establishing ethical standards for AI in ART presents a complex challenge due to the diverse cultural and moral viewpoints on reproductive technologies. Significant ethical issues include the opacity and interpretability of AI systems, which are frequently perceived as "black-box" models, as well as uncertainties regarding their generalisability.

To address these concerns, there are growing recommendations for enhancing the interpretability, explainability, and transparency of AI tools and data. Suggested approaches include developing unified registries where IVF clinics would contribute data and making both data and code publicly accessible (Afnan et al., 2022; Chow et al., 2021). Such measures would facilitate independent validation and replication of models while maintaining the confidentiality of patient and embryo data (Afnan et al., 2021). Prioritising transparency and reliability in AI programming is considered by many as crucial for mitigating hidden biases and enhancing the generalizability, fairness, and trustworthiness of AI-based healthcare solutions (Hanassab et al., 2024; Lee et al., 2024). However, the majority of models are not recoverable, with developers invoking ethics and privacy concerns to justify the fact that videos and other patient data are not publicly available nor redistributable (Duval et al., 2023).

From a social studies of science and technology perspective, the notion of transparency functions as a boundary object (Bowker & Star, 2000) in empirical research. It serves to align diverse social actors around a shared ethical principle, providing a common reference point across different contexts. At the same time, transparency is sufficiently flexible to accommodate the practical demands of achieving effectiveness in various settings.

Effectiveness and Clinical Validation

The field of AI is currently facing a number of significant challenges when considering the potential clinical utility of such technologies. The focus on the limited evidence for the effectiveness of AI in enhancing ART success rates, as well as the diverse methodologies employed in the validation of AI applications for ART in clinical settings, represent a pivotal stage in the transition from theoretical effectiveness to practical utility. The unproven nature of "add-on"[2] treatments offered by fertility clinics, often promoted within a "hope market" as described by Perrotta (2024), had already been analysed by social scientists before the advent of AI in ART (e.g., Becker, 2000; Franklin, 2022; Perrotta, 2024). In addition to concerns about private clinics profiting from treatments that impose financial burdens on vulnerable patients (Iacoponi et al., 2022; van de Wiel et al., 2020), discussions regarding AI in ART have increasingly centred on how to ensure validation of these technologies across different geographies (e.g., Afnan et al., 2021; Chow et al., 2021; Hanassab et al., 2024; Salih et al., 2023; Wilkinson et al., 2019).

In the view of some academic experts in the field of reproductive medicine and embryology, there is a discrepancy between the academic preference for rigorous evaluation of AI models through well-powered, high-quality randomised controlled trials (RCTs) and the industry tendency to favour larger, international, multi-centre, non-randomised

---

[2] The term "add-on" is subject to interpretation and may vary in its application depending on the context, time and geographic location. In general, it is used to describe "additional" treatments or procedures that are distinct from those already in place.

studies aimed at confirming early promising results, validating algorithms and training AI systems.

> Salient efforts from both academia and industry have validated the utility of retrospective data to enable data-driven decision-making for ART. To ensure viable deployment, these models can benefit from larger, multi-center datasets that incorporate both heterogeneous patient populations and also capture the idiosyncratic nature of clinical practice world-wide. (…) Prospective validation (e.g., well-designed RCTs) with relevant outcome measures is a key step to assess the efficacy and efficiency of these models in clinical environments and thus demonstrate impact on patient outcomes. (Hanassab et al., 2024, p. 16)

The divide between academia and industry highlights the existence of different priorities and methodologies, which are influenced by interpretive flexibility (Hess & Sovacool, 2020). This means that knowledge and technological design are shaped through social processes involving negotiation over their meaning. In the specific case of clinically validating AI in ART, there is a dynamic process of constructing, maintaining, contesting, and interpreting standards. In evidence-based medicine, randomised controlled trials are considered the gold standard due to their rigorous control of variables. However, randomised controlled trials are resource-intensive and may not always be feasible for emerging technologies from the industry's perspective. The industry often prefers non-randomised studies because they have broader applicability and yield faster results.

The significance of randomised controlled trials was reaffirmed in a 2019 opinion paper in Human Reproduction, a flagship journal of the European Society of Human Reproduction and Embryology (ESHRE). Written by academics from biostatistics and health sciences (Wilkinson et al., 2019), the paper addressed the emerging narrative suggesting the decline of RCTs in reproductive medicine. This narrative arose from hypothetical advances in prediction tools and algorithm-driven IVF, which were believed to support personalised treatments. The opinion piece, titled "Don't abandon RCTs in IVF. We don't even understand them", called for improved trials. The authors acknowledged the challenges of conducting trials in infertility research but stressed the need to advance randomised controlled trials. Despite existing methodological work, they emphasised the importance of collaboration between experts, including those with lived experience of infertility and IVF. To ensure well-powered, high-quality trials, the authors urged coordinated efforts. They highlighted the need for consistent reporting of core outcome measures and methods that facilitate evidence synthesis.

An analysis of the motivations behind different approaches to clinical validation reveals underlying power dynamics and stakeholder interests. The academic community's advocacy for high-quality randomised controlled trials reflects a commitment to scientific rigour and patient safety. This approach ensures that AI technologies are thoroughly vetted before widespread implementation, preventing the premature adoption of technologies based on incomplete evidence.

In contrast, the industry's preference for larger, non-randomised trials reflects a pragmatic approach that prioritises rapid deployment and market readiness. The focus on "teaching the computer" highlights the commercial imperative to refine and validate algorithms quickly to capitalise on early advances. The technical distinctions between randomised controlled trials and non-randomised studies thus map onto broader social and political distinctions between scientific rigour and pragmatism. These distinctions have significant social and political implications, and raise important ethical questions about the process of validating AI in ART. This is consistent with what science and technology studies refers to as the "politics of design", which examines the political and societal implications of design choices in technologies (Hess & Sovacool, 2020). For example, the drive for rapid validation must be carefully weighed against the responsibility to ensure that AI technologies do not exacerbate existing inequities or introduce new harms.

Technological Evolution

The prospective convergence of AI in ART with an unlimited supply of in vitro-developed gametes[3] has the potential to trigger a new wave of societal debates about the implications and future clinical applications of these technologies (Hengstschläger, 2023; Horta et al., 2023). The gametes in question can be derived from induced pluripotent stem cells, which are generated from parental skin cells. This process involves taking skin cells and "reprogramming" them into a pluripotent state, allowing them to develop into eggs and sperm. These cells have the potential to allow individuals to extend their fertility and create embryos from their own bodies, while overcoming the potential shortage of gametes or embryos for research purposes (Hendriks et al., 2015; Meskus, 2021; Smajdor, 2019; Zhang et al., 2020).

Given the potential implications for our understanding of human reproduction, it is essential to engage in a comprehensive ethical discussion about the convergence of AI in ART and artificial gametes, from the laboratory bench to clinical practice. As highlighted by Keating and Cambrosio (2003), Franklin (2013), and Wahlberg and colleagues (2021), the established conventions for sharing protocols, instruments, substances, and projects between laboratories offer significant opportunities for advancing technoscientific visions of future possibilities. This evolving landscape will not only redefine our comprehension of human reproduction but also shape the future possibilities in ART.

The Divide Between Developers and Critics

The following table provides a concise summary of contrasting views on AI in ART from two broadly defined social groups ("developers" and "critics"), without addressing their internal diversity(Table 3.3).

---

[3] The term "artificial gametes" is used to refer to somatic (non-reproductive) cells, stem cells, or progenitor cells that have been "reprogrammed" to function as eggs or sperm.

Table 3.3 Diverse views on primary clinical AI applications in assisted reproductive technologies

Applications[a]	Views of developers	Views of critics
Assessing patients' reproductive potential	• Improve efficiency in the estimation of the patient's overall change of success (e.g., evaluate ovarian reserve and endometrial receptivity; predict oocyte quality; predict semen profile by accessing person's lifestyle) • Provide digital home kits for semen analysis	• Potential for self-surveillance • Fear of reproductive surveillance and social sorting of patients • Concerns about: lack of clinical validation, generalisation, transparency; access and potential for exacerbate social inequalities; putative generation of misleading conclusions • Frameworks for validation and regulation are yet to be formalised
Monitoring treatment	• Improve treatment options and better planning of the procedures: optimise and personalise drug selection and dosing for ovarian stimulation, and induction of oocyte maturation; suggest the most individualised treatment of recurrent implantation failure; determine the optimal timing for egg retrieval and the best strategy for embryo transfer	• Concerns about: lack of clinical validation, generalisation, transparency, trust; access and potential for exacerbate social inequalities; whether adoption is of value for patients • Frameworks for validation and regulation (e.g., data sharing, privacy, informed consent) are yet to be formalised

(continued)

Table 3.3 (continued)

Applications[a]	Views of developers	Views of critics
Sperm/oocyte assessment	• Select the most competent eggs and sperm with speed and precision • Reduced potential for human error and intra-operator subjectivity and variability associated with manual assessment • Automated data recording	• Implicit moral implications in the selection of gametes • Concerns about: lack of clinical validation, generalisation, transparency, trust; access and potential for exacerbate social inequalities; whether adoption is of value for patients; putative generation of misleading conclusions • Frameworks for validation and regulation (e.g., data sharing, privacy, informed consent) are yet to be formalised
Embryo selection	• Analyse large amounts of complex data in real time to select the best embryos for transfer, automating some of the processes involved and reducing costs • Reduce potential for human error and intra-operator subjectivity and variability associated with manual assessment	• Implicit moral implications in the selection of embryos • Concerns about: lack of clinical validation, generalisation, transparency, trust; access and potential for exacerbate social inequalities; whether adoption is of value for patients; potential responsibility gap for AI choices • Frameworks for validation and regulation (e.g., data sharing, privacy, informed consent) are yet to be formalised

[a] The synthesis of the principal clinical applications of AI in ART is based on a review of the academic literature in the field of ART (Hanassab et al., 2024; Medenica et al., 2022; Zaninovic & Rosenwaks, 2020). Some surgical clinical AI applications in ART are not included in this table. It is the case of robotics-assisted surgery, used for microsurgical testicular sperm extraction, vasectomy reversal, myomectomy, and ovarian tissue cryopreservation, among others

Towards Acceptable and Feasible Artificial Intelligence for Embryo Selection

In this section, we examine the ways in which developers and experts supporting AI in ART collectively perform a work of mythologising (Sapir, 2020) to legitimise the commercialisation of innovative AI models in the field of embryo selection. The work of mythologising refers to the process by which the narrative of embryo selection is elevated to the status of myth within the field of reproductive medicine. The case of embryo selection is particularly illustrative of a process whereby a story acquires mythological status through synchronised processes of strategic publications in renowned journals and diffusion and reconstruction on clinics' websites.

We will describe ongoing efforts to promote specific ethical principles, establish standards, and combine real clinical data and synthetic data in the practices of knowledge production and adjudication. These efforts aim to reaffirm the legitimacy of AI development in response to multiple and diverse counter-narratives from academics, clinicians, and embryologists, who criticise, among other things, the lack of validation, generalizability, and transparency.

The driving force behind the development of AI applications in the embryology laboratory emerged in parallel with the exponential growth in computing power and the accumulation of embryo imaging data (Lee et al., 2024). This was coupled with the notion that one of the primary determinants of successful pregnancy is embryo quality. When considering investment in AI applications related to embryo selection, it is important to note that retrospective analysis of embryo-related data, such as videos, images, and clinical outcomes, has often been exempt from ethical review (e.g., Tran et al., 2019). Consequently, research on embryo selection by AI is underpinned by material-discursive techno-scientific practices that delineate boundaries between ethics directed at humans, embryos, and data related to embryos (videos, images, and clinical outcomes), always accompanied by exclusions and always open to contestation (Haraway, 1991; Meskus, 2021).

Promoting Reliability, Human-Centricity, and Accessibility

In a seminal publication in Human Reproduction,[4] a flagship journal of the European Society of Human Reproduction and Embryology (ESHRE), Tran et al. (2019) disseminated the creation of a deep learning[5] model named IVY. This model predicted the likelihood of a pregnancy with a foetal heart directly from raw time-lapse video of the embryos.[6] The researchers were affiliated with Harrison AI, a global healthcare technology company headquartered in Australia, and Virtus Health, one of the world's top five providers of assisted reproductive services, with a market-leading position in Australia, Ireland, and Denmark, and a growing presence in the UK and Singapore. They received grant support from Vitrolife Group, the manufacturer of the Embryoscope time-lapse imaging device used in the study to generate data to support deep learning. The authors emphasised the need for further studies, including the use of the gold standard for evaluating the efficacy and safety of medical interventions, namely randomised controlled trials. This plea served to validate the continued development and commercialisation of IVY, despite the existence of certain concerns, as evidenced by the website of IVFAustralia, the leading fertility provider in Australia and member of Virtus Health.

The potential benefits of automation and standardisation in embryo selection have been highlighted, despite the lack of clarity regarding the clinical impact and clinical significance of the proposed deep learning

---

[4] This paper is considered a seminal publication as it is the first research article found in a search conducted in the journals of the European Society of Human Reproduction and Embryology (ESHRE) and the American Society of Human Reproduction (ASRM), the leading societies for reproductive medicine and fertility worldwide. The search was conducted using "artificial intelligence" as keyword or topic (retrieved 21 June, 2024, from https://academic.oup.com/humrep and https://www.fertstert.org/).

[5] Deep learning is a subfield of machine learning that is based on learning hierarchical knowledge from data rather than rule-based programming (Tran et al., 2019, p. 1012).

[6] These videos, which are short, allow the evolution of embryos to be revealed and followed over five days in a watchable sequence. The technique involves the sequential capture of images or video frames at predefined intervals, followed by the playback of these frames at a higher speed. This technique creates the illusion of time moving faster, thereby revealing patterns and processes that are otherwise too slow to observe directly by embryologists.

model. Given these uncertainties, it is appropriate to ask how the developers of IVY are responding to them.

IVFAustralia is currently using IVY. According to their website, this organisation was the first group in Australia to develop an AI system to analyse embryo growth in order to predict the greatest likelihood of a viable pregnancy resulting from the transfer of a single embryo. A video is embedded in the website of IVFAustralia that demonstrates how AI can "help the embryologist to select the best embryo for transfer" (retrieved 21 June, 2024, from https://www.youtube.com/watch?v= WQIySjtidQE). This video brings the concept of AI tangible and serves as an effective communication tool for developers, embryologists, reproductive specialists, and patients. Moreover, the objective is to alleviate concerns regarding the lack of safety and potential machine determinism. To achieve this, a "human-in-the-loop" approach is used, meaning that AI assists embryologists in their decision-making process, but the final decision is always made by the embryologist.

The visions of the use of AI in embryo selection presented on the IVFAustralia website function as an interface between the present and the future, playing a crucial role in the practices and social arrangements that its developers believe are necessary for the widespread adoption of AI in IVF treatments. The future widespread use of AI in embryo selection depends on the "rigorous evaluation of the technology" through an ongoing randomised controlled trial overseen by IVF Australia scientists. The need for a randomised controlled trial was previously announced in the aforementioned paper, published five years ago (Tran et al., 2019). On the website of IVFAustralia, the following announcement was displayed:

> IVFAustralia is carrying out a rigorous assessment of the technology through a randomised controlled trial of artificial intelligence in embryo selection. As the pioneers of this groundbreaking technology for Virtus Health, IVFAustralia's scientists will continue to oversee the project, taking it across all our Virtus sites. We anticipate that artificial intelligence will be rolling out across our IVFAustralia sites following the trial

as part of our ongoing investment in the very highest standards of laboratory technology. (Retrieved 21 June, 2024, from https://www.ivf.com.au/success-rates/our-laboratories/ivy-artificial-intelligence-in-ivf)

Given the seemingly innocuous nature of this announcement, it is important to examine a warning issued by academics in biostatistics and health sciences in an opinion paper published in Human Reproduction. This paper was intended to address an emerging narrative advocating the end of randomised controlled trials in reproductive medicine (Wilkinson et al., 2019). As these critical academics have noted, many clinics rely on the willingness of patients to quickly accept innovative treatments in reproductive medicine, despite a lack of evidence to support their effectiveness and efficacy. This raises questions about the use of AI-based ART in a way that is scientifically rigorous, ethically sound, and socially responsible.

> The field of reproductive medicine is a hotbed of innovation. Patients understandably want to give themselves the best possible chance of having a baby, and many clinics trade on this fact to sell add-on interventions with claims of treatment superiority. In this environment, a new treatment can quickly become entrenched if a scientific study appears to support its use. (Wilkinson et al., 2019, p. 2093)

An alternative AI-based assessment model for the prediction of embryo viability was published in April 2020 also in the journal Human Reproduction (VerMilyea et al., 2020). This model, known as Life Whisperer AI, was developed by Ovation Fertility—a leading IVF provider in the United States—in collaboration with Presagen, a global AI technology company specialising in healthcare products. The researchers argued that existing methods using time-lapse videos to predict successful pregnancies were both costly and unreliable. Instead, they claimed their model, which analyses single images of Day 5 embryos taken with conventional microscopes, was more effective. They also introduced a cloud-based software application designed to make this model accessible and scalable for IVF clinics worldwide, presenting it as a cost-effective and widely applicable solution.

The Life Whisperer AI model and the IVY deep learning model are examples of the development of different AI-based methods to improve embryo selection. All have been characterised as "objective, standardised and efficient tools" (Zaninovic & Rosenwaks, 2020, p. 914). The mobilisation of the myth of objectivity, standardisation, and efficiency in the face of different AI-based methods for the same purpose reveals the interpretive flexibility of developers' standards. Interpretive flexibility recognises that the perspectives, interests, and contexts of different social actors shape the way in which they interpret and apply standards in a variety of ways (Hess & Sovacool, 2020). Although standards are understood as rules that establish uniformities across space and time and are used to guide the implementation of AI in ART, the case of embryo selection demonstrates how the implementation and understanding of standardised tools can vary according to the specific needs, goals, and ethical considerations of different groups of developers.

Setting Up Standards

It is important to consider the broader implications of standards beyond their seemingly benign role. Standards can act as a form of indirect governance for AI, effectively serving as a mechanism to circumvent or delay formal legal oversight (Solow-Niederman, 2023). Furthermore, they have the capacity to redistribute traditional regulatory functions from public bodies to privatised and business-oriented entities (Rommetveit & Van Dijk, 2022). While AI standards may appear to provide a neutral and objective framework for navigating a complex and contested normative landscape (Solow-Niederman, 2023), in practice their effectiveness can be compromised by political dynamics. These standards are influenced by actors making normative choices within particular institutional contexts, which are often shaped by political and economic incentives and constraints.

One of the incentives for developers and experts supporting AI in ART to engage in the construction and development of standards in the 2020s was the anticipated expansion of the use of AI in reproductive medicine. This anticipated expansion depended on the development of standards to overcome the challenges associated with the use of different methods and algorithms on different platforms. The expected expansion of AI in the field of ART would occur not only through the application of AI tools to a vast amount of embryological, clinical, and genetic data, but also through the implementation of AI technologies in all aspects of fertility treatments. The developers of AI applications for ART have presented it as a universal solution that is capable of ensuring the delivery of personalised treatments:

> The ultimate goal will be to apply AI tools to the analysis of all embryological, clinical, and genetic data in an effort to provide patient-tailored treatments (…). The potential that AI will usher in a new era of standardization, automatization, and precision to IVF has engendered wide enthusiasm and has even gained traction in the commercial arena. While applications of AI in embryology have gained the most attention and shown great promise, it is likely its use will widen to other aspects of reproductive medicine. (Zaninovic & Rosenwaks, 2020, p. 914 and p. 919)

Despite the great expectations surrounding the potential of AI in clinical settings, developers have observed a "slow uptake" of AI in reproductive medicine, attributing this "delay" to concerns among clinicians and embryologists regarding the use of different AI techniques in daily clinic routine. Furthermore, there was a lack of in-depth interdisciplinary communication and uncertainty regarding the generalisation of AI systems for different populations (Fernandez et al., 2020; Fraire-Zamora et al., 2022; Sadeghi, 2022). This perception draws attention to the ways in which assisted reproductive service providers, as users, have often engaged with AI in pragmatic ways, as active consumers who question, resist, reject, and reinterpret the applications of AI in ART (Geampana & Perrotta, 2023; Hess & Sovacool, 2020).

In this context, the call for standards can be understood as a technical solution to address the concerns of reproductive specialists, who were primarily represented by the developers as a workforce in need of training in interdisciplinary skills, critical thinking, and technical expertise in order to use AI in ART as expected by the developers (Hanassab et al., 2024; Serdarogullari et al., 2022). The construction of standards is thus mobilised by AI developers as a means of fostering support and trust among practitioners and embryologists. This approach can be seen as an attempt to co-opt critical insights from sociological and STS literature on public trust as part of efforts to integrate AI solutions into clinical and laboratory workflows. However, this strategy has been criticised for reassuring practices and cautious approaches that might otherwise hinder the developers' agenda (Machado et al., 2023).

In conclusion, expectations around standardisation played a crucial role in embryo selection, coexisting in tension with local algorithmic practices (Geampana & Perrotta, 2023). These standards—technical, political, and discursive in nature—shaped socio-ethical debates and mitigated some concerns around the use of AI in ART. As Bowker and Star, scholars in the Social Studies of Science and Technology, illustrate through their analysis of standardisation processes, standards typically operate behind the scenes of scientific and technical practices, serving as an invisible part of the infrastructure when functioning effectively. However, they become visible when there are failures or disruptions in this infrastructure (Bowker & Star, 2000). In the context of AI in ART, the importance of standards was influenced by reproductive specialists' concerns about uncertainties that had the potential to "delay" the expected progress of AI in reproductive medicine. The push for further standardisation helped to maintain social order in the development and commercialisation of AI systems, sidelining the social and ethical negotiations involved in the creation and implementation of these standards.

Combining Real Clinical Data and Synthetic Data

To date, only in-house personalised AI algorithms have been developed in the field of embryo selection, as a result of complex human-technology interactions, and therefore vary from clinic to clinic (Geampana & Perrotta, 2023). This suggests that the generalisability of AI models is limited when applied to a different region or ethnic group (Salih et al., 2023). The potential for algorithms to generalise poorly to different populations perpetuates bias (Afnan et al., 2021). From the perspective of developers and those supporting AI in ART, limitations in the availability of training data, privacy concerns, and the potential for algorithmic bias can be addressed by combining real clinical data (clinics' own customised database of local patient populations) and synthetic data inputs (data generated by AI, ostensibly to increase the diversity of datasets and protect privacy). The combination of real clinical data with synthetic data has been shielded from ethical scrutiny, particularly in the field of ART, despite significant concerns about its utility, privacy guarantees, and regulatory oversight (Jordan et al., 2022).

As sociologist Benjamin Jacobsen (2023) shows, developers' promises place algorithms beyond the realm of risk and actively reconfigure the conditions of possibility for machine learning in contemporary society. This is illustrated by the case of embryo selection. The separation of the ethical and political implications of synthetic training data for machine-learning algorithms from real data related to embryos and patients has led to the sociotechnical imaginary that social problems, power, and ethics can be more broadly excluded from AI (Helm et al., 2024). Moreover, as James Steinhoff (2024) critically observes in his article "Toward a political economy of synthetic data", the question of whether data-intensive capitalism is a form of surveillance capitalism has been raised by critical scholars. The significant investment in synthetic data by data-intensive companies supports the autonomisation of the circuit of data-intensive capital, independent of the oversight of human subjects. In conclusion, these concerns demonstrate how the combination of synthetic data with real clinical data for model training and validation may introduce new controversies and uncertainties with epistemological, ontological, and political-economic consequences that urgently require critical attention.

Future Avenues of Artificial Intelligence in Assisted Reproductive Technologies

Emerging and controversial AI applications for ART continue to develop and promise to open the door to a new era of human reproduction. This has been described as "the next frontier in the journey towards personalised reproductive medicine and improved fertility outcomes for patients" (Chow et al., 2021, p. C33). A number of AI-driven startups have developed AI-based solutions that promise to provide personalised fertility treatment plans. This new era is accompanied by a sociotechnical imaginary that AI is inherently objective, purely data-driven, fast, accurate, and relentlessly efficient. Here are some of the common myths about AI that have been mobilised by developers and experts supporting AI techniques:

> The capacity of AI techniques to analyze large amounts of complex data such as images and timelapse objectively, whereby non-invasive assessment of gametes and embryos can be done in real-time, has significant potential for future impact in achieving healthy live birth. (Hanassab et al., 2024, p. 7)

The visions of "commercial innovation" promoted by the private sector create numerous potential pathways for exploration, driven by the cultural myth that AI is an enduring presence that should be universally applied. This includes examining various innovative approaches where technology intersects with bodily intimacy (Haraway, 1991), enabling processes of self-surveillance and self-awareness (Hampshire & Simpson, 2015; Law, 2020; Myers & Martin, 2021), and distributing the responsibility for the success of ART among various social actors through the lens of AI.

This process of distributing responsibility for the success of ART is exemplified by the technologies announced by Aniket Tapre, CEO of Multiply, a US-based IVF-tech startup. These technologies included the "lab on a chip" (end-to-end IVF, ICSI, biopsy, and embryo freezing in a box) and "DIY" [do-it-yourself] IVF cycles (retrieved 21 June, 2024, from https://www.mpo-mag.com/contents/view_online-exclusives/2021-12-09/ai-based-platforms-are-shaping-the-ivf-industry), among others. The former is an integrated system for the entire IVF process, while the latter is a "self-service IVF cycle".

While DIY IVF cycles may reduce treatment costs and make AI in ART more accessible and convenient to the wider public, they significantly reduce human interaction and raise additional concerns about data security and privacy, access, consent, and validation in different populations, as well as the potential for expanded reproductive surveillance. As AI in ART moves from the embryology laboratory to IVF clinics and the marketplace, controversies and uncertainties emerge, inspiring further instances of resistance and reinterpretation beyond the "powerful ecosystem" that drives its deployment:

> As technological advances move from the research lab to the IVF clinic and marketplace, and as companies form to pursue commercial applications, we see a number of similarities in how and why they are being developed—and a powerful ecosystem is taking shape to drive their development. (Brayboy & Quaas, 2023, p. 262)

In a recent opinion paper, academics Kiri Beilby and Karin Hammarberg (2024) suggest that healthcare providers should consider investing in generating their own models using the emerging technology of ChatGPT in the near future. The potential impact of generative AI, such as ChatGPT, on informed fertility-related decision-making is a cause for concern. This is due to the fact that the information used to train ChatGPT is not publicly known, the lack of source attribution and incorrect attribution, and the potential to provide recommendations based on inaccurate or untrustworthy information (Beilby & Hammarberg, 2024). Furthermore, summarising digital content that is uncertain or controversial could exclude some views and silence the diversity of positions on the primary applications of AI in ART explored in this chapter.

Networks and Ethical Assemblages Around Artificial Intelligence in Assisted Reproductive Technologies

In this section we summarise the complex and often divergent positions of different social actors involved in AI-based ART, and how these intertwine with other elements (technologies, organisations, values, ideas, and practices) to shape unstable and evolving collective understandings of what is considered ethical and "good practices" in relation to AI in ART. These ethical assemblages, as discussed in Chapter 1 and explored in the case of facial recognition technologies in Chapter 2, emerge and take shape through evolving networks. In the case of AI in ART, these networks involve relationships between different social actors from academia, industry, and clinical practice in reproductive medicine and fertility, each with their own political and economic interests, logics, and power relations (Bigo et al., 2019). Table 3.4 provides a systematisation of the networks and the different social actors and domain-specific hierarchical premises involved in AI in ART, as well as the related sociomaterial and symbolic landscapes analysed in the previous sections of this chapter.

There are four ways in which different views of the controversies and uncertainties surrounding AI in ART are intertwined to create and continually negotiate and transform the ethical debate around these technologies: embryo selection; personalisation of fertility treatments; relationships between academia and industry; and self-regulation. First, by prioritising embryo selection as the gateway to enhancing the role of AI in reproductive medicine and improving ART success rates, it tends to establish AI as a dominant influence. The expansion of AI in ART as a practice capable of influencing ethical debates is staged as a balancing act between the promise of efficiency and personalised patient care and controversial or unknown issues related to ethical boundaries, access and inequality, effectiveness and clinical validation, meaningful outcomes, transparency, and trustworthiness.

Second, the advocacy of personalised reproductive medicine as a means to improve success rates, streamline processes, and broaden access to parenthood aligns AI in ART with the interests and needs of both patients and healthcare providers. However, proponents of personalised ART often neglect or downplay concerns and controversies that are shaped by the values and priorities of academic scientists and AI healthcare companies. This notion of personalised ART thus plays an important role in ethical debates involving different actors, negotiating a delicate balance between establishing ethical standards and best practices, ensuring informed consent and decision-making, and addressing

Table 3.4 Networks and ethical assemblages around AI in assisted reproductive technologies

Actors	Networks	Ethical assemblages
Academic scientists	• Conduct fundamental and applied research in AI and ART • Engage in collaborations with other academic institutions and industry partners • Participate in academic and professional organisations, influencing standards and practices in the field • Publish critical analysis	• Adhere to ethical guidelines and codes of conduct set by academic bodies to address issues related to clinical validation of AI-driven ART • Advocate for transparency and generalisability in AI development processes • Engage with industry to discuss ethical standards based on scientific evidence and research • Shape the ethical discourse and call for regulatory oversight
Healthcare technology companies	• Develop and deploy AI-driven ART • Control data necessary for training algorithms • Shape market availability and adoption • Economic power over clinics • Strategic publications in renowned journals	• Conduct retrospective data analysis often under scrutiny of their own company's ethical boards • Need support and trust from clinicians, embryologists, and patients • Interpret and apply standards in a variety of ways to facilitated commercialisation • Privacy and pragmatism often lead to support "black-box" AI

(continued)

Table 3.4 (continued)

Actors	Networks	Ethical assemblages
Healthcare providers	• Use, question, resist, refuse, and reinterpret the applications of AI to make daily decisions • Influence how AI is integrated into laboratory and clinical workflows • Provide databases for research and development • Participate in professional societies, influencing standards and practices in the field	• Collaborate with tech companies and research institutions • Raise awareness on the need to enhance the clinical relevance and utility of AI-driven ART • Leading professional societies influence agenda setting, ethical guidelines, and good practices
Patients	• Provide data, often without knowing, used in AI-driven ART • Use AI-driven ART, but their experiences and positions on ethical issues are not known	• Mobilised by academic scientists to demand transparency, meaningful outcomes such as a healthy live birth, and ethical standards • Mobilised by some clinics to sell add-on interventions with claims of treatment superiority

the uncertainties and controversies surrounding the clinical validation, relevance and utility of AI in ART.

Third, it is important to recognise the hybrid roles of scientists working on AI in ART. Academics often focus on critically analysing the ethical implications of these technologies, advocating clinical validation through randomised controlled trials, and calling for strong ethical standards and regulatory oversight. In contrast, scientists working in industry publish strategic articles in prestigious journals to gain support for the development and commercialisation of AI applications, even when their clinical impact and significance is not always clear. These applications are often based on retrospective data analysis and are reviewed by their own corporate ethics committees. In addition, there is often interpretive flexibility in ethical standards, such as varying interpretations of transparency guidelines. Developers and some academics may use these ambiguities to justify the use of "black-box" AI, while still promoting best practice in line with fundamental ethical principles. This interpretive flexibility is particularly evident in the varying interpretations of transparency guidelines, which may change depending on the practical outcomes of black-box AI models.

Fourth, the deployment of AI in ART on the periphery of traditional regulatory frameworks shifts the discourse on ethical issues to privatised assisted reproductive services and business-oriented health technology companies. This shift brings academics and healthcare providers into the discussion, but largely excludes other publics. This technocratic, expert-driven approach emphasises reliability, human-centredness, and accessibility as key ethical standards for the use of AI in ART, while producing nuanced and hybrid classifications of transparency, consent, informed choice, privacy, and data protection. Moreover, self-regulation within these private and business-oriented entities often shapes and defines these ethical standards, further entrenching a system that may overlook broader societal concerns.

Conclusion

The concept of ethical assemblages reveals how controversies and uncertainties about AI in ART are framed and articulated. These processes are deeply entangled with symbolic, social, and material orders, including political and economic agendas and visions. Assemblages highlight how ethical issues are part of dynamic and interconnected networks involving different social actors from academia, industry, and clinical practice in reproductive medicine and fertility. Together, these actors generate diverse and complex social and technological transformations that evolve over time.

The advent of data-driven ART, in which data and computers configure decision-making linked to visions of desirable futures for human reproduction and ART, as well as good scientific and clinical practices, requires a new framework for ethics in line with the concept of ethical assemblages. This framework facilitates an understanding of the context and meaning of human actions and their impact on life, identity, institutions, and society. It also allows for the critical examination and shaping of social structures and power dynamics (Boenig-Liptsin, 2022).

In this sense, it is important to pay more attention to the management of conflicts of interest, values, and priorities in clinical and research settings (Duffy et al., 2021) and their links to entrepreneurial activity (McNeil et al., 2017). As reproductive medicine in the context of ART begins to understand itself and its problems through the paradigm of AI, a machine-learning medical-political order (Amoore, 2023) emerges that seeks to optimise abstract representations of data, potentially decoupling AI in ART from its tangible and corporeal, often complex impacts on people and societies in real-life scenarios.

In addition, further research is needed to understand how basic, clinical, and translational research on AI in ART and the use of these technologies by patients and healthcare providers form ethical assemblages that may potentially alter, displace, or render redundant the concept of "embodied progress" (Franklin, 2022) in data-driven societies. In Sarah Franklin's perspective, "embodied progress" refers to the notion that technological advancements, particularly in the field of ART, should translate into tangible improvements in human well-being and experiences. This concept emphasises the direct impact of technological innovations on the body and lived experience, suggesting that progress should not only be measured by technical achievements but also by their meaningful enhancement of human life. To explore how ethical assemblages of AI in the field of ART can reshape and challenge our understandings of the body, subjectivity, ethics, and social relations, potentially displacing or redefining what progress means in practice in data-driven ART, is crucial to understand the kind of society that AI in ART can bring into being.

Ethical discussions have raised significant concerns about the social inequalities that AI in ART could exacerbate. These conversations also question whether these technologies will truly benefit patients and improve success rates. In addition, there is discourse about the potential for misleading conclusions and limited generalisability of AI applications. Such debates highlight the risks and challenges associated with AI in ART and advocate for rigorous clinical validation in diverse populations, improved data sharing, and optimisation of algorithms while pursuing technological advancement. Finally, the increasing use of AI to analyse extensive embryological, clinical, and genetic data, and its integration into all aspects of fertility treatment, has raised significant concerns about transparency, privacy, consent, and decision-making, particularly in the absence of fully established regulatory frameworks.

This chapter has explored a range of positions on the current controversies and uncertainties surrounding the future trajectories of AI in ART in the context of debates that have been dominated by experts and technocratic approaches within networks that include mainly business-oriented health technology companies, privatised assisted reproductive service providers, and some academics and health care providers. A greater hybridity and boundary fusion of these networks could facilitate the expansion of collective imaginations and spaces of possibility regarding the potential for ethical assemblages to operate differently in a variety of social and political contexts, involving heterogeneous networks that interweave different social actors, institutions, and social values.

References

Afnan, M., Afnan, M. A. M., Liu, Y., Savulescu, J., Mishra, A., Conitzer, V., & Rudin, C. (2022). Data solidarity for machine learning for embryo selection: A call for the creation of an open access repository of embryo data. *Reproductive BioMedicine Online, 45*(1), 10–13. https://doi.org/10.1016/j.rbmo.2022.03.015

Afnan, M. A. M., Liu, Y., Conitzer, V., Rudin, C., Mishra, A., & Savulescu, J. (2021). Interpretable, not black-box, artificial intelligence should be used for embryo selection. *Human Reproduction Open, 4*, hoab040. https://doi.org/10.1093/hropen/hoab040

Amoore, L. (2023). Machine learning political orders. *Review of International Studies, 49*(1), 20–36. https://doi.org/10.1017/S0260210522000031

Amoore, L. (2020). *Cloud ethics*. Duke University Press.

Appadurai, A. (2013). *The future as cultural fact: Essays on the global condition.* Verso.

Aradau, C., & Blanke, T. (2022). Algorithmic reason: The new government of self and others. *Oxford University Press.* https://doi.org/10.1093/oso/978019 2859624.001.0001

Bamford, T., Easter, C., Montgomery, S., Smith, R., Dhillon-Smith, R. K., Barrie, A., Campbell, A., & Coomarasamy, A. (2023). A comparison of 12 machine learning models developed to predict ploidy, using a morphokinetic meta-dataset of 8147 embryos. *Human Reproduction, 38*(4), 569–581. https://doi.org/10.1093/humrep/dead034

Bazzani, G. (2023). Futures in action: Expectations, imaginaries and narratives of the future. *Sociology, 57*(2), 382–397. https://doi.org/10.1177/003803 85221138010

Becker, G. (2000). *The elusive embryo: How women and men approach new reproductive technologies.* University of California Press.

Beilby, K., & Hammarberg, K. (2024). ChatGPT: A reliable fertility decision-making tool? *Human Reproduction, 39*(3), 443–447. https://doi.org/10.1093/humrep/dead272

Bigo, D., Isin, E., & Ruppert, E. (2019). *Data politics. Worlds, subjects, rights.* Routledge.

Bowker, G. C., & Star, S. L. (2000). *Sorting things out: Classification and its consequences.* The MIT Press.

Boenig-Liptsin, M. (2022). Aiming at the good life in the datafied world: A co-productionist framework of ethics. *Big Data and Society, 9*(2). https://doi.org/10.1177/20539517221139782

Brayboy, L. M., & Quaas, A. M. (2023). The DIY IVF cycle—Harnessing the power of deeptech to bring ART to the masses. *Journal of Assisted Reproduction and Genetics, 40*(2), 259–263. https://doi.org/10.1007/s10815-022-02691-x

Chow, D. J. X., Wijesinghe, P., Dholakia, K., & Dunning, K. L. (2021). Does artificial intelligence have a role in the IVF clinic? *Reproduction and Fertility, 2*(3), C29–C34. https://doi.org/10.1530/RAF-21-0043

Duffy, J. M. N., Adamson, G. D., Benson, E., Bhattacharya, S., Bofill, M., Brian, K., Collura, B., Curtis, C., Evers, J. L. H., Farquharson, R. G., Fincham, A., Franik, S., Giudice, L. C., Glanville, E., Hickey, M., Horne, A. W., Hull, M. L., Johnson, N. P., Jordan, V., ... Youssef, M. A. (2021). Top 10 priorities for future infertility research: An international consensus development study. *Fertility and Sterility, 115*(1), 180–190. https://doi.org/10.1016/j.fertnstert.2020.11.014

Duval, A., Nogueira, D., Dissler, N., Maskani Filali, M., Delestro Matos, F., Chansel-Debordeaux, L., Ferrer-Buitrago, M., Ferrer, E., Antequera, V., Ruiz-Jorro, M., Papaxanthos, A., Ouchchane, H., Keppi, B., Prima, P. Y., Regnier-Vigouroux, G., Trebesses, L., Geoffroy-Siraudin, C., Zaragoza, S., Scalici, E., … Boussommier-Calleja, A. (2023). A hybrid artificial intelligence model leverages multi-centric clinical data to improve fetal heart rate pregnancy prediction across time-lapse systems. *Human Reproduction, 38*(4), 596–608. https://doi.org/10.1093/humrep/dead023

Fernandez, E. I., Ferreira, A. S., Cecílio, M. H. M., Chéles, D. S., de Souza, R. C. M., Nogueira, M. F. G., & Rocha, J. C. (2020). Artificial intelligence in the IVF laboratory: Overview through the application of different types of algorithms for the classification of reproductive data. *Journal of Assisted Reproduction and Genetics, 37*, 2359–2376. https://doi.org/10.1007/s10815-020-01881-9

Ferrand, T., Boulant, J., He, C., Chambost, J., Jacques, C., Pena, C. A., Hickman, C., Reignier, A., & Fréour, T. (2023). Predicting the number of oocytes retrieved from controlled ovarian hyperstimulation with machine learning. *Human Reproduction, 38*(10), 1918–1926. https://doi.org/10.1093/humrep/dead163

Fraire-Zamora, J. J., Ali, Z. E., Makieva, S., Massarotti, C., Kohlhepp, F., Liperis, G., Perugini, M., Thambawita, V., & Mincheva, M. (2022). #ESHREjc report: On the road to preconception and personalized counselling with machine learning models. *Human Reproduction, 37*(8), 1955–1957. https://doi.org/10.1093/humrep/deac111

Franklin, S. (2013). *Biological relatives: IVF, stem cells, and the future of kinship.* Duke University Press.

Franklin, S. (2022). *Embodied progress: A cultural account of assisted conception* (2nd ed.). Routledge.

Geampana, A., & Perrotta, M. (2023). Predicting success in the embryology lab: The use of algorithmic technologies in knowledge production. *Science, Technology and Human Values, 48*(1), 212–233. https://doi.org/10.1177/01622439211057105

Hampshire, K., & Simpson, B. (Eds.). (2015). *Assisted reproductive technologies in the third phase: global encounters and emerging moral worlds.* Berghahn Books.

Hanassab, S., Abbara, A., Yeung, A. C., Voliotis, M., Tsaneva-Atanasova, K., Kelsey, T. W., Trew, G. H., Nelson, S. M., Heinis, T., & Dhillo, W. S.

(2024). The prospect of artificial intelligence to personalize assisted reproductive technology. *Npj Digital Medicine, 7*(55). https://doi.org/10.1038/s41746-024-01006-x

Haraway, D. (1988). Situated knowledges: The science question in feminism and the privilege of partial perspective. *Feminist Studies, 14*(3), 575–599. http://www.jstor.org/stable/3178066

Haraway, D. (1991). *Simians, cyborgs, and women. The reinvention of nature.* Routledge.

Helm, P., Lipp, B., & Pujadas, R. (2024). Generating reality and silencing debate: Synthetic data as discursive device. *Big Data and Society, 11*(2). https://doi.org/10.1177/20539517241249447

Hendriks, S., Dancet, E. A., van Pelt, A. M., Hamer, G., & Repping, S. (2015). Artificial gametes: A systematic review of biological progress towards clinical application. *Human Reproduction Update, 21*(3), 285–296. https://doi.org/10.1093/humupd/dmv001

Hengstschläger, M. (2023). Artificial intelligence as a door opener for a new era of human reproduction. *Human Reproduction Open, 2023*(4). https://doi.org/10.1093/hropen/hoad043

Hess, D. J., & Sovacool, B. K. (2020). Sociotechnical matters: Reviewing and integrating science and technology studies with energy social science. *Energy Research and Social Science, 65*(January), 101462. https://doi.org/10.1016/j.erss.2020.101462

Hogarth, S. (2017). Valley of the unicorns: Consumer genomics, venture capital and digital disruption. *New Genetics and Society, 36*(3), 250–272. https://doi.org/10.1080/14636778.2017.1352469

Horta, F., Salih, M., Austin, C., Warty, R., Smith, V., Rolnik, D. L., Reddy, S., Rezatofighi, H., & Vollenhoven, B. (2023). Reply: Artificial intelligence as a door opener for a new era of human reproduction. *Human Reproduction Open, 2023*(4). https://doi.org/10.1093/hropen/hoad045

Iacoponi, O., van de Wiel, L., Wilkinson, J., & Harper, J. C. (2022). Passion, pressure and pragmatism: How fertility clinic medical directors view IVF add-ons. *Reproductive BioMedicine Online, 45*(1), 169–179. https://doi.org/10.1016/j.rbmo.2022.02.021

Jacobsen, B. N. (2023). Machine learning and the politics of synthetic data. *Big Data and Society, 10*(1). https://doi.org/10.1177/20539517221145372

Jordon, J., Szpruch, L., Houssiau, F., Bottarelli, M., Cherubin, G., Maple, C., Cohen, S. N., & Weller, A. (2022). *Synthetic data—What, why and how?* The Alan Turing Institute. http://arxiv.org/abs/2205.03257

Keating, P., & Cambrosio, A. (2003). *Biomedical platforms: Realigning the normal and the pathological in late-twentieth-century medicine.* MIT Press.

Law, C. (2020). Biologically infallible? Men's views on male age-related fertility decline and sperm freezing. *Sociology of Health & Illness, 42*(6), 1409–1423. https://doi.org/10.1111/1467-9566.13116

Lee, T., Natalwala, J., Chapple, V., & Liu, Y. (2024). A brief history of artificial intelligence embryo selection: From black-box to glass-box. *Human Reproduction, 39*(2), 285–292. https://doi.org/10.1093/humrep/dead254

Machado, H., Silva, S., & Neiva, L. (2023). Publics' views on ethical challenges of artificial intelligence: A scoping review. *AI Ethics.* https://doi.org/10.1007/s43681-023-00387-1

Martin, K. (2019). Ethical implications and accountability of algorithms. *Journal of Business Ethics, 160*(4), 835–850. https://doi.org/10.1007/s10551-018-3921-3

McNeil, M., Arribas-Ayllon, M., Haran, J., Mackenzie, A., & Tutton, R. (2017). Conceptualizing imaginaries of science, technology and society. In U. Felt, R. Fouche, C. A. Miller, & L. Smith-Doerr (Eds.), *The handbook of science and technology studies* (4th ed., pp. 435–464). MIT Press.

Medenica, S., Zivanovic, D., Batkoska, L., Marinelli, S., Basile, G., Perino, A., Cucinella, G., Gullo, G., & Zaami, S. (2022). The future is coming: Artificial intelligence in the treatment of infertility could improve assisted reproduction outcomes—The value of regulatory frameworks. *Diagnostics, 12*, 2979. https://doi.org/10.3390/diagnostics12122979

Meskus, M. (2021). Speculative feminism and the shifting frontiers of bioscience: Envisioning reproductive futures with synthetic gametes through the ethnographic method. *Feminist Theory, 24*(2), 151–169. https://doi.org/10.1177/14647001211030174

Myers, K. C., & Martin, L. J. (2021). Freezing time? The sociology of egg freezing. *Sociology Compass, 15*(4), e12850. https://doi.org/10.1111/soc4.12850

Patrizio, P., Albertini, D. F., Gleicher, N., & Caplan, A. (2022). The changing world of IVF: The pros and cons of new business models offering assisted reproductive technologies. *Journal of Assisted Reproduction and Genetics, 39*(2), 305–313. https://doi.org/10.1007/s10815-022-02399-y

Perrotta, M. (2024). *Biomedical innovation in fertility care: Evidence challenges, commercialization, and the market for hope.* Bristol University Press. https://doi.org/10.51952/9781529236750

Riegler, M. A., Stensen, M. H., Witczak, O., Andersen, J. M., Hicks, S. A., Hammer, H. L., Delbarre, E., Halvorsen, P., Yazidi, A., Holst, N., &

Haugen, T. B. (2021). Artificial intelligence in the fertility clinic: Status, pitfalls and possibilities. *Human Reproduction, 36*(9), 2429–2442. https://doi.org/10.1093/humrep/deab168

Rommetveit, K., & van Dijk, N. (2022). Privacy engineering and the techno-regulatory imaginary. *Social Studies of Science, 52*(6), 853–877. https://doi.org/10.1177/03063127221111942

Rubel, A., Castro, C., & Pham, A. (2019). Agency laundering and information technologies. *Ethical Theory and Moral Practice, 22*(4), 1017–1041. https://doi.org/10.1007/s10677-019-10030-w

Sadeghi, M. R. (2022). Will artificial intelligence change the future of IVF? *Journal of Reproduction and Infertility, 23*(3), 139–140. https://doi.org/10.18502/jri.v23i3.10003

Salih, M., Austin, C., Warty, R. R., Tiktin, C., Rolnik, D. L., Momeni, M., Rezatofighi, H., Reddy, S., Smith, V., Vollenhoven, B., & Horta, F. (2023). Embryo selection through artificial intelligence versus embryologists: a systematic review. *Human Reproduction Open, 2023*(3). https://doi.org/10.1093/hropen/hoad031

Sapir, A. (2020). Mythologizing the story of a scientific invention: Constructing the legitimacy of research commercialization. *Organization Studies, 41*(6), 799–820. https://doi.org/10.1177/0170840618814575

Serdarogullari, M., Ammar, O. F., Sharma, K., Kohlhepp, F., Montjean, D., Meseguer, M., & Fraire-Zamora, J. J. (2022). #ESHREjc report: Seeing is believing! How time lapse imaging can improve IVF practice and take it to the future clinic. *Human Reproduction, 37*(6), 1370–1372. https://doi.org/10.1093/humrep/deac072

Silva, S., & Machado, H. (2011). The construction of meaning by experts and would-be parents in assisted reproductive technology. *Sociology of Health & Illness, 33*(6), 853–868. https://doi.org/10.1111/j.1467-9566.2010.01327.x

Smajdor, A. (2019). An alternative to sexual reproduction: Artificial gametes and their implications for society. *British Medical Bulletin, 129*(1), 79–89. https://doi.org/10.1093/bmb/ldz001

Solow-Niederman, A. (2023). Can AI standards have politics? *UCLA Law Review, 2*(April 2023), 2–17. https://ssrn.com/abstract=4714812

Steinhoff, J. (2024). Toward a political economy of synthetic data: A data-intensive capitalism that is not a surveillance capitalism? *New Media and Society, 26*(6), 3290–3306. https://doi.org/10.1177/14614448221099217

Tran, D., Cooke, S., Illingworth, P. J., & Gardner, D. K. (2019). Deep learning as a predictive tool for fetal heart pregnancy following time-lapse incubation and blastocyst transfer. *Human Reproduction, 34*(6), 1011–1018. https://doi.org/10.1093/humrep/dez064

Tsamados, A., Aggarwal, N., Cowls, J., Morley, J., Roberts, H., Taddeo, M., & Floridi, L. (2022). The ethics of algorithms: Key problems and solutions. *AI and Society, 37*(1), 215–230. https://doi.org/10.1007/s00146-021-01154-8

van de Wiel, L., Wilkinson, J., Athanasiou, P., & Harper, J. (2020). The prevalence, promotion and pricing of three IVF add-ons on fertility clinic websites. *Reproductive BioMedicine Online, 41*(5), 801–806. https://doi.org/10.1016/j.rbmo.2020.07.021

VerMilyea, M., Hall, J. M. M., Diakiw, S. M., Johnston, A., Nguyen, T., Perugini, D., Miller, A., Picou, A., Murphy, A. P., & Perugini, M. (2020). Development of an artificial intelligence-based assessment model for prediction of embryo viability using static images captured by optical light microscopy during IVF. *Human Reproduction, 35*(4), 770–784. https://doi.org/10.1093/HUMREP/DEAA013

Wahlberg, A., Dong, D., Song, P., & Jianfeng, Z. (2021). The platforming of human embryo editing: Prospecting "disease free" futures. *New Genetics and Society, 40*(4), 367–383. https://doi.org/10.1080/14636778.2021.1997578

Wilkinson, J., Brison, D. R., Duffy, J. M. N., Farquhar, C. M., Lensen, S., Mastenbroek, S., Van Wely, M., & Vail, A. (2019). Don't abandon RCTs in IVF. We don't even understand them. *Human Reproduction, 34*(11), 2093–2098. https://doi.org/10.1093/humrep/dez199

Zaninovic, N., & Rosenwaks, Z. (2020). Artificial intelligence in human in vitro fertilization and embryology. *Fertility & Sterility, 114*(5), 914–920. https://doi.org/10.1016/j.fertnstert.2020.09.157

Zhang, P. Y., Fan, Y., Tan, T., & Yu, Y. (2020). Generation of artificial gamete and embryo from stem cells in reproductive medicine. *Frontiers in Bioengineering and Biotechnology, 8*, 781. https://doi.org/10.3389/fbioe.2020.00781

4

Conclusion

Abstract The conclusion of this book emphasises the critical necessity of understanding artificial intelligence (AI) through the lens of ethical assemblages. By examining cases such as facial recognition technologies and assisted reproductive technologies, this concluding chapter elucidates the complex and evolving networks of interactions that shape the ethical landscape of AI. It demonstrates how ethical debates, shaped by societal values, play a pivotal role in the innovation processes, influencing both the development and adaptation of AI technologies. The conclusion highlights the need to reassess socio-material and symbolic landscapes in order to understand the profound impacts of AI on individual rights and societal norms. This reassessment is crucial for fostering a digital society in which technological progress is aligned with democratic values and human rights. The book urges for further research into the evolving ethical controversies and uncertainties surrounding AI, particularly how these debates shape public perceptions and policy-making over time.

Keywords Ethical assemblages · Artificial intelligence · AI ethics · Facial recognition technologies · Assisted reproductive technologies

© The Author(s), under exclusive license to Springer Nature   **115**
Singapore Pte Ltd. 2025
H. Machado and S. Silva, *Ethical Assemblages of Artificial Intelligence*,
https://doi.org/10.1007/978-981-96-4158-1_4

This book offers a reflection on artificial intelligence (AI) through the lens of ethical assemblages, drawing on concrete examples from the fields of facial recognition technologies and assisted reproductive technologies (ART). It demonstrates how diverse, situated, and contextual processes of negotiating and discussing ethical issues unfold in different parts of the world. The strength of this conceptual framework lies in its ability to reveal the complex, dynamic, fluid, and evolving networks of interactions between different elements—technologies, people, organisations, values, ideas, and practices—that collectively shape the development, deployment, and impact of multiple and unstable versions of AI technologies.

The contributions of the ethical assemblages approach include a deeper exploration of the inherent societal values and relationships conveyed in ethical debates about AI, and how these social processes are shaped by the expectations embedded in ideas about innovation. Our analysis considers how these dynamics allocate advantages and disadvantages, and regulate what is permitted or prohibited, revealing how the multiple controversies and uncertainties surrounding AI paradoxically serve as the underlying conditions and driving forces behind increasingly ambitious innovation efforts. This perspective offers a critical vantage point for understanding the complex interplay between the diverse nexus and logics of social actors and domain-specific premises of hierarchisation, technological progress, and the imperative to navigate the ethical landscape beyond the widespread implicit assumption that AI ethics can serve as a comprehensive solution to controversies and uncertainties. As such, this book highlights the need to explore ethical frameworks in order to better navigate the challenges of AI.

This book synthesises the insights of ethical assemblages of AI gained from the cases of facial recognition technologies and ART. It demonstrates that the varying stages of the innovation cycle compel developers, academics, and professional groups to utilise AI tools in their work. This, in turn, modifies the framing of the benefits and potential harms of technologies to address ethical concerns and align with societal values—from the initial stages of conception and development to the subsequent stages of adaptation and redirection. This process does not necessarily impede the core mechanisms that drive innovation. Instead, it provides a means to negotiate novelty or legitimacy, and to refine business and scientific practices to address immediate concerns, while still allowing the essential engines of technological progress to operate. The arguments mobilised in negotiating and contesting ethical issues could act as incentives to increase data collection efforts and to refine and improve algorithms, while maintaining the fundamental imperatives of technological progress. To delve into the ethical assemblages of AI in facial recognition technologies and ART, we provide a concise summary in Table 4.1, juxtaposing the divergences and the blurred boundaries between these two technologies.

Ethical controversies and uncertainties surrounding these technologies develop within complex networks characterised by data dependency, widespread use, and interdisciplinary collaboration, where different forms of power, knowledge, and contestation interact. Both facial recognition technologies and ART are part of broader innovation ecosystems that include academic research, commercial development, regulatory frameworks, and public discourses, and rely heavily on large data sets to function effectively. In addition, the development and implementation of both technologies involve collaboration across multiple disciplines, including computer science, ethics, medicine, and social sciences.

However, these two technologies differ significantly in the type of data they use, their societal impact, and the instances of contestation. Facial recognition technologies primarily deal with large image databases and visual data to train algorithms for facial identification, whereas AI in ART uses morphological, clinical, and genetic data from images to improve clinical outcomes. Over the past two decades, facial recognition technologies have been employed extensively in a multitude of contexts across society, demonstrating an evolving blurring of boundaries between acceptable and unacceptable applications. This has occurred concurrently with the emergence of significant legal and ethical concerns pertaining to privacy and civil liberties. Facial recognition technologies are frequently linked with matters of security, safety, and surveillance. They have a plethora of applications, ranging from law enforcement to personal device security. However, they typically involve the non-consensual collection of data in public spaces, which gives rise to ethical

Table 4.1 Ethical assemblages: Controversies, uncertainties, and networks

Blurred boundaries	Divergences between	
	Facial recognition technologies	Assisted reproductive technologies
Elements of the networks (technologies, people, organisations, values, ideas, practices)		
Data dependency	Extensive image databases and visual data	Large, temporal, and dynamic datasets generated throughout ART cycles, comprising images, morphological, clinical, and genetic information
Widespread usages; increase data collection efforts	From national security context to integration into everyday life for the last two decades; blurring of boundaries between acceptable and unacceptable uses	From embryo selection to all aspects of IVF treatments; more recent applications primarily targeting individuals seeking fertility treatment
Interdisciplinary collaboration, where different forms of power, knowledge, and contestation interact	Individual and social movements of resistance, refusal, and dissensus	Professional dissensus and refusal
Academic scientists and technology companies	Security agencies, privacy advocates: regulators and policymakers, governments, non-governmental organizations, general citizens	Healthcare providers and patients
Advocate human-centricity, generalisability, trust, engaging with publics	Public security, safety, and surveillance	Improve the success rates of ART, personalisation of care

(continued)

Table 4.1 (continued)

Blurred boundaries	Divergences between	
	Facial recognition technologies	Assisted reproductive technologies
Optimisation (efficiency, effectiveness, political realm)	Pervasive, affecting public spaces, personal freedom, and mass surveillance practices	Private spaces, affecting individual lives, future generations, and clinical and laboratory workflows and reproductive decisions
Controversies and uncertainties		
Inconsistent practices regarding consent and privacy	Widespread and often unauthorized surveillance by governments or corporations, threat to personal privacy and autonomy	To understand, explain, and consent to the use of AI is challenging, raising concerns about informed and shared decision-making, and potential misrepresentation of patients' values and preferences
Exacerbate social inequalities and reinforce discrimination; apprehensions about reliability across different populations	- Racial, gender, age, and disability biases in algorithmic decision-making - Potential misuse and abuse for targeting marginalized communities and minorities	- Concerns about the morality of the algorithms - Lack of clinical validation and effectiveness - Debates around what constitutes meaningful outcomes
Lack of transparency	How/what data is being collected and managed	Concerns about machine paternalistic decision-making process
Lack of clear regulations and accountability	Insufficient regulation and oversight of public-private partnerships	Potential responsibility gap and agency laundering for decisions made by AI

(continued)

Table 4.1 (continued)

Blurred boundaries	Divergences between	
	Facial recognition technologies	Assisted reproductive technologies
Use of personal data for profit	Use of data for commercial purposes, without individuals' informed consent	Concerns about equitable access and benefits
Rapid and speculative advances and convergence of different technologies	Continuous adaptation of ethical guidelines and regulations, uncertainty about future developments and their implications	Emergence of new social pressures for using AI in ART; implications of potential future convergence with in vitro gametes
Ethical assemblages		
Academic research - Pursuit of knowledge, innovation, and advancement of technology - Distinction between good and bad practices	- Advocate strict ethical standards - Critical analysis of ethical and social issues - Concerns with ethical bad practices from some scientists and industry: debates around assessment of data collection and distribution methods	- Research collaborative networks (data sharing and collaboration with fertility clinics) to enhance transparency, reliability, and generalisability - Advocate rigorous validation through randomised controlled trials, - Publish critical analysis

(continued)

Table 4.1 (continued)

Blurred boundaries	Divergences between	
	Facial recognition technologies	Assisted reproductive technologies
Commercial development - Profit and competitive advantage and innovation - Corporate social responsibility and strategic techno-solutionism: future-oriented enactments of rights and ethical principles integrated into technological frameworks	- Control data - Shape market availability and adoption - Eternal optimisation of algorithms - Distribution of responsibility for privacy and data governance	- Control information over clinics - Future-oriented promises strategically published in renowned journals and disseminated on clinics' websites - Interpretive flexibility of standards - Combine real clinical data and synthetic data - Focus on larger international multi-centre non-randomised studies

(continued)

Table 4.1 (continued)

Blurred boundaries	Divergences between	
	Facial recognition technologies	Assisted reproductive technologies
Regulatory frameworks - Balance innovation with human rights and public good	Essential tools of policing for governments *versus* activist resistance of non-governmental organisations and regulators against racism, discrimination, and infringement on individuals' rights and civil liberties	- Self-regulation - Inconsistency regarding levels of de-identification of patient information and adequate informed consent
Public discourses - Translation between current societal problems and visions of the future, affecting market adoption - Influence regulatory policies and scrutinize practices	Nuanced attitudes - Personal privacy and security + public safety - Concerns with unfair and intrusive surveillance; security concerns about potential abuses and malicious use - Value the offer of tangible public benefits, regulation and oversight to prevent the misuse and to protect citizens' rights and freedoms	- Mobilization of patients' interests, but their experiences and positions are not known - Emerging controversies and uncertainties inspire further instances of contestation

concerns surrounding unauthorised surveillance. The societal implications of facial recognition technologies can be far-reaching, affecting public spaces, personal freedoms, and mass surveillance practices.

In contrast, AI in ART is a more recent entrant that aims to improve success rates and personalised care, circulating mainly in private spaces and primarily targeting individuals seeking fertility treatment. The incorporation of AI into clinical and laboratory workflows and reproductive decisions has a more profound impact on individual lives, with the potential to influence the genetic traits of future generations through the implicit morality of algorithms. This raises ethical concerns about the limits of intervention in human reproduction, access and inequality, informed decision-making, efficacy, and clinical validation.

There are different forms of contestation of AI technologies from a wide range of regions of the world. These range from activist resistance by NGOs and regulators concerned with human rights and oppression, scientists concerned with ethical malpractice by other scientists and industry, citizens concerned with unfair surveillance, to overt and multi-pronged resistance strategies orchestrated by technology companies, to policy circles and some media calling for oversight to prevent the misuse of facial recognition technologies in surveillance. The contestation of facial recognition technologies may manifest in various forms, including direct opposition to their authoritarian imposition and legal challenges to their ethical implications. Furthermore, contestation may manifest in more nuanced forms, such as individuals concealing their identities from facial recognition systems or declining to utilise AI applications in the context of ART. Dissensus, in this context, represents a form of active dissent, whereby stakeholders challenge the underlying norms and assumptions that drive the adoption of these technologies, thereby disrupting the appearance of consensual acceptance and underscoring the necessity for a broader societal discourse.

Ethical controversies and uncertainties involve a variety of social actors who form distinct yet interwoven networks. These networks include academics, technology companies, regulators, and ordinary people. The ethical debates surrounding facial recognition technologies are significantly shaped by the input of various stakeholders, including security agencies, technology companies, and privacy advocates (such as global multi-stakeholder platforms, non-governmental organisations (NGOs), and civil society organisations). Conversely, the integration of artificial intelligence (AI) into ART practices is largely influenced by private assisted reproductive service providers, health technology companies, and professionals (such as clinicians and embryologists). Each of the social actors that comprise the networks has its own interests and concerns regarding facial recognition technologies and ART. They are aligned in their advocacy for optimisation, human-centricity, generalisability of AI applications, trust in the use of AI, and engagement with the public. However, there is divergence in terms of the ethical use, consent, and governance of these technologies.

Academic scientists and technology companies are shaping the ethical debate around AI in facial recognition technologies and ART. Academic scientists are primarily driven by the pursuit of knowledge, innovation, and the advancement of technology. They manage their reputations and protect their authority and the credibility of science by maintaining strict ethical standards and research collaborative networks, publishing critical analyses, and constructing boundaries between good and bad science (e.g., in relation to their relationships with industry and other scientists who may not adhere to ethical guidelines, or the lack of randomised control trials). Technology companies prioritise profitability and are driven by competitive advantage and innovation in products and services (e.g., optimising algorithms), often published in prestigious journals or disseminated on their websites. Many technology companies also recognise the importance of corporate social responsibility, including the ethical use of technology and addressing societal concerns (e.g., by sharing responsibility and engaging with other stakeholders), essentially as a reputation strategy. These companies focus on strategic techno-solutionism, based on forward-looking enactments of rights and ethical principles built into technological frameworks (e.g., through eternal optimisation of algorithms, rapidly validated by large international multi-centre non-randomised studies, interpretive flexibility of standards, and combination of real and synthetic data). Nevertheless, there are notable differences between the involvement of various social actors in the development and implementation of facial recognition technologies and ART. In the case of facial recognition technologies, general citizens, NGOs, regulators and policymakers, and governments play a pivotal role in the ethical considerations surrounding these technologies. This contrasts with the involvement of healthcare providers and patients in the ethical considerations pertaining to AI in ART.

The general public attaches great importance to the protection of their personal privacy and security, and is concerned about the manner in which their data is collected, stored, and used. Moreover, they emphasise the importance of public safety, regulation, and oversight in order to ensure that technological advancement is balanced with the protection of citizens' rights and freedoms. NGOs, meanwhile, often prioritise the protection of human rights, advocating for justice and combating racism and discrimination. It is their objective to ensure that technologies such as facial recognition technologies do not infringe upon individuals' freedoms and privacy. Furthermore, they also advocate for transparency in the utilisation of technologies and accountability for any misuse or ethical violations.

The objective of regulators and policymakers, on the other hand, is to establish and implement regulations that guarantee the ethical utilisation of facial recognition technologies, striking a balance between innovation and societal benefit. They are guided by the public interest, with the aim of protecting citizens from potential harms while facilitating beneficial technological advancements. Governments, however, place a premium on national security, law enforcement, and public safety, viewing facial recognition technologies as a means of crime prevention and efficient public service delivery. Other stakeholders may regard state surveillance as an infringement of civil rights and liberties.

In healthcare, providers employ, question, decline to utilise, and reinterpret the applications of AI to inform their daily decisions, thereby influencing the manner in which AI is integrated into laboratory and clinical workflows. They also advocate for the clinical relevance and utility of AI applications. Patients' interests are mobilised by academic scientists to demand transparency, meaningful outcomes, and ethical standards, and some private clinics to sell innovative treatments. However, their experiences and positions surrounding the use of AI in ART remain unknown.

The question of how to balance innovation with robust ethical oversight to ensure that AI technologies serve the public good without infringing on individual rights or societal values is the subject of ongoing debate and diverse positionings. The following controversies and uncertainties are situated at the margins of the networks of both facial recognition technologies and ART: privacy and consent; bias, errors, shortcomings, and inaccuracies; transparency issues; regulation and oversight; commercial exploitation; and technological evolution. The normalisation of extensive usage of AI technologies, which become unquestioned, even expected and demanded, remains outside the scope of ethical discourse, or on the fringes, and obscures the underlying power dynamics and social, material, and symbolic logics of systemic biases.

- *Privacy and consent*: Facial recognition technologies have the potential to identify, monitor, and track individuals extensively without their consent, thereby posing a threat to personal privacy and autonomy. Furthermore, they can undermine the right to control one's own data and may result in widespread unauthorised surveillance by governments or corporations. In some countries, there have been collaborative initiatives between law enforcement agencies and private companies to implement facial recognition technology in public spaces, including shopping centres and football stadiums. With regard to the application of AI in ART, the intricate and opaque nature of AI systems can render it challenging for prospective parents, practitioners, and embryologists to gain comprehensive insight, elucidate the intricacies, and provide informed consent to the utilisation of such systems. This raises concerns pertaining to the tenets of informed and shared decision-making, and the potential misrepresentation of their values and preferences.
- *Bias, errors, shortcomings, and inaccuracies*: Both technologies are vulnerable to the influence of biases and implicit moralities embedded within their algorithms, which may give rise to discriminatory practices and outcomes that serve to reinforce existing stereotypes. For example, facial recognition technologies frequently demonstrate elevated error rates for specific ethnic groups. There is a risk of facial recognition technologies being misused and abused to target marginalised communities. The accuracy of both technologies varies considerably, depending on factors such as demographic diversity. Consequently, there is uncertainty regarding their reliability across different populations. This raises concerns about the lack of validation and effectiveness of AI in ART, as well as debates around what should be considered meaningful outcomes.
- *Transparency issues*: The decision-making processes of both technologies may operate without sufficient transparency, which makes it challenging for individuals to ascertain how AI technologies are being utilised, the data being collected, and the manner in which the data is being managed and applied.
- *Regulation and oversight*: The lack of clear regulations governing the use of both technologies frequently results in inconsistent practices

and a dearth of accountability for how the technologies are used, particularly in the case of public-private partnerships in facial recognition technologies. The regulatory landscape is still evolving, with varying guidelines and recommendations across different countries, which may create responsibility gaps and uncertainty for developers and users regarding compliance and legal liabilities.

- *Commercial exploitation*: Private companies might utilise personal data for financial gain. In the context of facial recognition technologies, data may be employed for commercial purposes, such as targeted advertising, without the informed consent of individuals. The commercial exploitation of ART gives rise to concerns regarding equitable access to reproductive healthcare and equitable benefits derived from AI applications, which are predominantly accessible to wealthier individuals and large private clinics.
- *Technological evolution*: The rapid and speculative advances in both technologies mean that ethical guidelines and regulations must continuously adapt, creating uncertainty about future developments and their implications. The emergence of social pressures for using AI in ART might inspire further instances of contestation, as there is a lack of consensus on the ethical implications of using AI in this context.

In conclusion, an understanding of the ethical assemblages of AI requires a fundamental reassessment of the socio-material and symbolic landscapes involved. Such a reassessment must take into account the profound implications of these technologies for matters pertaining to surveillance and care, citizenship, activism, and social movements. In consequence, a more nuanced understanding of the impact of AI technologies on individual rights and societal norms may be attained. Ultimately, such an understanding can facilitate the promotion of a fairer and more just digital society, in which technological advances serve to enhance, rather than undermine, democratic values and human rights.

Further in-depth studies on the intricate networks related to AI technologies, involving both insiders and outsiders, would be beneficial. Furthermore, research is required on how distinct phases within innovation cycles impact the framing of technologies in terms of their benefits and potential harms, specifically how these framings influence public opinion and policy-making. Finally, there is an opportunity to investigate how ethical controversies and uncertainties evolve over time and what their long-term effects are.

Index

GPSR Compliance
The European Union's (EU) General Product Safety Regulation (GPSR) is a set
of rules that requires consumer products to be safe and our obligations to
ensure this.

If you have any concerns about our products, you can contact us on

ProductSafety@springernature.com

In case Publisher is established outside the EU, the EU authorized
representative is:

Springer Nature Customer Service Center GmbH
Europaplatz 3
69115 Heidelberg, Germany